# USS LEXINGTON

by Steve Wiper

## CLASSIC WARSHIPS PUBLISHING

P. O. Box 57591 • Tucson, AZ. 85732 • USA
Web Site: www.classicwarships.com • Telephone (520)748-2992
Copyright © October 2009
**ISBN 978-0-9823583-3-7**
Printed by Arizona Lithographers, Tucson, Arizona

# GENERAL HISTORY OF THE USS LEXINGTON CV-2

The primary focus of this book will be of USS LEXINGTON CV-2 during her operations in the opening months of the Pacific War and her eventual loss. A brief coverage of her pre-war history will be offered in both text and photography. For statistical information about LEXINGTON please refer to the sources listed at the back of this book.

The fourth LEXINGTON (CV-2), originally designated CC-1, was laid down as a battle cruiser, 8 January 1921 by Fore River Shipbuilding Co., Quincy, Massachusetts. Authorized to be completed as an aircraft carrier, 1 July 1922, she was launched 3 October 1925, and commissioned 14 December 1927.

After fitting out and shakedown, LEXINGTON joined the battle fleet at San Pedro, California, 7 April 1928. Based there, she operated on the west coast with Aircraft Squadrons, Battle Fleet, in flight training, tactical exercises, and battle problems. Each year she participated in fleet maneuvers off the Hawaiian Islands, the Caribbean Sea, off the Panama Canal Zone, and in eastern Pacific Ocean.

LEXINGTON received her magnetic degaussing system during a refit in March-April 1941. This consisted of massive cables attached to the hull sides externally. These wrapped around the entire length of the hull, at about the hangar deck level, and was quite visible in photographs of her in her final days of 1941, through to her demise in May 1942. She also had a YE aircraft homing beacon installed about this time. It's very prominent antenna was mounted atop the foremast.

During a refit in May 1941, at the Puget Sound Navy Yard, LEXINGTON had her crash barriers on the flight deck relocated, and life netting fitted along the lengths of the flight deck edges, port and starboard. The forward machine gun platforms at the flight deck level were each extended and fitted with a single 3 in./50 cal. AA mount as well as fitting that mount type to the aft platforms, both port and starboard. On the bridge, the top of the deck house was altered. A pair of .50 cal. machine guns were fitted, port and starboard, and atop each 8in. turret for a total of eight mounts. After the refit, she proceeded to sea and steamed with the battle force to the Hawaiian Islands for tactical exercises, based at Pearl Harbor.

Sometime in late June, or July 1941, LEXINGTON was at the Pearl Harbor Navy Yard for another refit, at which time she received an CXAM-1 air search radar system. The antenna was installed similar to that of her sistership SARATOGA, at the top, forward edge, of the funnel. A funnel cap, only on the forward uptake, was fitted to keep funnel gases and whistle steam from interfering with the CXAM-1 antenna. The control room for that radar was installed on the forward face of the funnel, in place of the the of the secondary conning and primary flight control stations.

At some time during late August 1941, CV-2 was at the Puget Sound Navy Yard for a refit. It is believed that at sometime during this period her 3 in. AA mounts were each replaced with quadruple 1.1 in. AA mounts, five in total. It must be noted that all of these 1.1in AA mounts were not installed with power drives and were manually operated. Also, no directors were available for installation at that time. A catwalk was fitted from the back of the bridge to the CXAM-1 radar control station on the leading face of the funnel. She was also at the Hunter's Point Navy Yard for dry docking about 15 October 1941.

Task Force 12 and LEXINGTON departed Pearl Harbor 5 December, en route to Midway. On 7 December 1941, LEXINGTON, at sea with TF-12, cruisers CHICAGO, PORTLAND, ASTORIA and five destroyers, carrying Marine Scout Bomber Squadron 231 from Pearl Harbor to reinforce Midway Island, when word of the Japanese attack on Pearl Harbor was received. They were about 425 miles to the southeast of Midway when this news was received. She immediately launched search planes to hunt for the Japanese fleet, and at midmorning headed south to rendezvous with the cruiser INDIANAPOLIS and later with the ENTERPRISE task force, TF-8, with heavy cruisers and destroyers, to conduct a search southwest of Oahu until returning to Pearl Harbor, 13 December 1941.

LEXINGTON sailed next day, 14 December, after refueling, as part of TF-11, with the cruisers INDIANAPOLIS, CHICAGO, PORTLAND, Destroyer Division 1, and fleet oiler NEOSHO. They were ordered to raid Japanese forces on Jaluit, in the Marshall Islands, to relieve pressure on Wake, as a diversionary raid. On the morning of 16 December, scout planes from LEXINGTON spotted a "RYUJO" class carrier 95 mile south of the TF. The oiler NEOSHO was sent west while LEXINGTON launched aircraft for a strike. The "carrier" turned out to be a derelict barge. On 17 December, LEXINGTON's cruiser escort conducted anti-aircraft practice, upon which they discovered that all their 5in. ammunition was faulty.

Orders to attack Jaluit were canceled 20 December, and TF-11, was directed to cover TF-14, consisting of CV-2's sistership SARATOGA, cruisers MINNEAPOLIS, ASTORIA, SAN FRANCISCO, Destroyer Squadron 4, aircraft depot ship TANGIER and the oiler NECHES, in reinforcing Wake. LEXINGTON's TF-12 was headed towards Wake Island from the south as a cover force for TF-14, when they joined TF-8, lead by carrier ENTERPRISE, cruisers NORTHAMPTON, CHESTER, SALT LAKE CITY, Destroyer Squadron 6, and the fleet oiler PLATTE, who were headed towards Wake from the east.

At 0911hrs. 22 December, all of the US Navy carrier task forces were ordered to break off the operation to relieve the US Marines on Wake Island, as the Japanese invasion had overrun the island fortress by that time. LEXINGTON's TF-11, along with the other carrier TFs set course for Pearl Harbor by 1400 hrs. 23 December, as all US forces on Wake had surrendered, arriving at Pearl Harbor on 27 December 1941.

LEXINGTON re-fueled, re-supplied and performed minor repairs for two days, then departed Pearl Harbor, 30 December 1941, and patrolled to block possible enemy raids in the Oahu-Johnston-Palmyra Islands triangle. She departed with only 17 F2As, one of which crashed on the stern of the carrier, ending up in the sea, the pilot rescued, thereby bringing her VF2 count down to 16 fighters. Her total air group numbered 16 F2As, 33 SBDs and 15 TBDs. Returning to Pearl Harbor, 3 January 1942, again re-fueling, re-supplying, but with an extended stay for crew rest, and a major electrical problem with one of her main electric drive motors that required an extended repair, she remained at Pearl Harbor until 7 January.

TF-11 with LEXINGTON departed Pearl Harbor, 7 January, to patrol northeast of Johnston Island. On 10 January, a patrol of F2As spotted a Japanese submarine running on the surface, about 60 miles south of TF-11. The sub crashed dived, but CV-2 later dispatched

F2As and TBDs with depth charges to the location, which spotted the sub again on the surface. The F2As lead the attack by strafing the sub with their .50cal MGs and the TBDs dropped their depth charges, as the sub crash dived again. The aircraft circled for about an hour, but the Japanese sub did not surface. It is believed that the sub was I-19, which was recorded as arriving at Kwajalein, 15 January. TF-11 returned to Pearl Harbor 16 January 1942.

LEXINGTON then departed Pearl Harbor, again as flagship for TF-11, supported by cruisers, MINNEAPOLIS, INDIANAPOLIS, PENSACOLA, SAN FRANCISCO, Destroyer Squadron 1, and fleet oiler NECHES, 22 January, to begin operations against Japanese forces at Wake Island. Steaming about 135 miles west of Oahu, Hawaii, 23 January 1942, NECHES was torpedoed by I-72 and sank mid-day. With the necessity to refuel during such a long range operation and then having been deprived of their oiler, TF-11 was forced to cancel the operation and return to Pearl Harbor. SARATOGA had been torpedoed just 12 days earlier, southwest of Oahu by a submarine from the same IJN 3rd. Submarine Division. TF-11 arrived at Pearl Harbor, 25 January 1942.

On 26 January 1942, VF-2 and it's 16 F2A Brewster Buffalo fighters were transferred to Marine Air Group 21 on Ford Island. On 30 January, 12 F4F Wildcat fighters were transferred to LEXINGTON, but these were from VF-3. While SARATOGA was laid up for torpedo damage repairs, her fighters nested aboard the LEXINGTON.

LEXINGTON and TF-11 departed Pearl Harbor, 31 January 1942. Their purpose was to cover eight troop transports with 20,000 troops which were reinforcing garrisons on Christmas, Canton and New Caledonia Islands. During this operation, TF-11 was temporarily joined by the cruisers QUINCY and VINCENNES, who were part of the escort for the transports from the US East Coast, transiting the Panama Canal. This was accomplished by 15 February, with the empty transports, QUINCY and VINCENNES, all departed for Pearl Harbor. TF-11 separated from the troop convoy and was then directed to operate under direction of the Australia-New Zealand Allied Command (ANZAC) Force.

On 16 February, the force headed north-west from the New Hebrides in order to carry out a carrier based raid upon Rabaul, New Britain. While approaching their air strike launch destination to the east of New Britain, 20 February, LEXINGTON's radar picked up a "bandit." Six F4F fighters were dispatched from the combat air patrol (CAP) and soon discovered three IJN four engined flying boats, shooting down two, the third escaping. Later that afternoon, CV-2's radar again picked up multiple "bandits," launching an additional six F4Fs, just as the enemy aircraft closed in on TF-11.

The first engagement of the Pacific War between United States Navy and Imperial Japanese Navy carrier based aircraft had begun. TF-11, more specifically LEXINGTON, was attacked by two waves of enemy aircraft, nine planes per wave. The action began at 1620 hrs., and continued intermittently until just after 1800 hrs. The carrier's own CAP and anti-aircraft (AA) fire was believed to have shot down 17 of the attackers. During a single sortie Lt. "Butch" O'Hare won the Medal of Honor by shooting down five Japanese Nakajima B5N2 "Kate" torpedo bombers with his Grumman F4F "Wildcat" fighter. As it turned out, the AA fire from LEXINGTON and her escorts in TF-11, was wildly inaccurate and was more a danger to CV-2's own CAP aircraft. There was strong evidence that all IJN aircraft downed in this action was by USN aircraft and not the AA fire from the warships of TF-11. The element of surprise was lost, and due to the fact that at that time, the USN then had just three carriers in the Pacific theater, and could not afford to lose any, the next day's planed attack upon Rabaul was cancelled and TF-11 turned back to the south.

By 24 February, LEXINGTON and TF-11 turned west and headed into the Coral Sea to hunt for suspected IJN vessels. Her offensive patrols in the Coral Sea continued until 6 March, when she rendezvoused west of the New Hebrides with TF-17. Together they were organized as CTF-11 (Combined), comprised of carriers LEXINGTON, YORKTOWN, cruisers MINNEAPOLIS, SAN FRANCISCO, INDIANAPOLIS, PENSACOLA, ASTORIA, LOUISVILLE, CHICAGO, HMAS AUSTRALIA, 14 destroyers and two fleet oilers. They proceeded west, back into the Coral Sea and headed north-west towards the Gulf of Papua, off southern New Guinea.

Initially, the plan was to launch a carrier based attack upon Rabaul, but just as CTF-11 arrived in the Gulf of Papua on 10 March, the plan changed to attack Salamaua and Lae, located on the eastern New Guinea coast, on Huon Gulf. Japanese forces had just landed there on 7 March and they were completely unaware of the approaching Allied task force. The task force was split, with the carriers, four cruisers and ten destroyers continuing north to their launching point and the two fleet oilers, four cruisers and four destroyers steaming to a rendezvous point to the south. The carriers then launched their aircraft for a thoroughly successful surprise attack flown over the 7500 ft. Owen Stanley Mountains of New Guinea to inflict moderate damage on shipping and installations at Salamaua and Lae, 10 March 1942. Only one US aircraft was lost in this raid. Although this was not a crucial raid, it was a huge boost in moral for a depressed navy and was also a important operation in the experience that was gained. CTF-11 departed the area to avoid any reprisal attacks from Japanese aircraft based at Rabaul. They later joined their fleet oilers to refuel in the central Coral Sea, on 14 March, then continued east towards the New Hebrides. LEXINGTON's TF-11 detached from YORKTOWN and TF-17, 16 March, south of the New Hebrides Islands and proceeded toward Pearl Harbor, arriving 26 March 1942.

TF-11 had just performed the longest war cruise in the US Navy with 54 days continuous operation in a very hostile war zone. LEXINGTON and the other vessels of TF-11 would spend the next three weeks at Pearl Harbor to perform repairs and maintenance to those warships. CV-2 was immediately taken in-hand by navy yard personnel for major modifications to her armament. Her 8 in. gun turrets were removed over a four day period and in their place a total of seven quadruple 1.1 in AA mounts were installed. A total of 22 of the new 20 mm AA machine guns were also fitted at the starboard base of the funnel, at the fore and aft ends of the ship, and in hinged platforms projecting from the boat wells in the hull side. LEXINGTON was also re-painted into a new camouflage measure, #11, using the new Navy Blue paint. Painting begun while the 8 in. turrets were being removed, but by the end of the first week of April, the massive aircraft carrier was completely painted overall a dark blue-gray.

On 14 April, she hoisted 14 F2A-3 fighters of VMF-211, US Marine Corp fighters aboard to ferry them to a Central Pacific island. These were the same fighters that the F4F Wildcat fighters of VF-3 had replaced back at the end of January 1942. CV-2 also received additional

F4F fighters about this time and they were all re-designated to VF-2.

LEXINGTON's TF-11, with the cruisers NEW ORLEANS, MINNEAPOLIS and 7 destroyers, sortied from Pearl Harbor 15 April 1942, headed south, by south-west, towards New Hebrides. At that time, CV-2's Air Group consisted of 21 F4F fighters, 37 SBD dive bombers and 12 TBD torpedo bombers, plus the 14 F2A fighters she was ferrying.

By 18 April, TF-11 steamed near enough to Palmyra Atoll, of the Line Islands spanning the Equator, Christmas Island being the most infamous, to launch the 14 F2A Buffalo fighters of VMF-211 to that destination. That accomplished, LEXINGTON and TF-11 steamed to rendezvous with the old battleships of TF-1 for battle training. This was cancelled late that night due to Naval Intelligence having information about a Japanese move towards the South Pacific. TF-11 was ordered to steam to a point 200 miles to the north of the Fiji Islands.

En route to that location, LEXINGTON's VF-2 launched for their first flight of the cruise, with 17 aircraft taking off on 19 April 1942. While the fighters of VF-2 were aloft on their training mission, LEXINGTON rendezvoused with the fleet oiler USS KASKASKIA AO-27. The next day, as CV-2 crossed the Equator, the crew performed the *"Crossing the Line"* ceremony, the seasoned *"Sheellbacks"* hazing the *"Pollywogs"* down the length of the flight deck, to go before and be judged by *"King Neptune and his Court."* Also, during that day, VF-2 flew an exercise with TBD Devastators of VT-2.

On 21 April 1942, orders came from the Commander in Chief, Pacific Fleet, for TF-11 to rendezvous with TF-17 about 250 miles northwest of New Caledonia on 1 May 1942. As TF-11, with LEXINGTON steamed in a southwesterly direction, VF-2 took up air patrols. On 25 April, they rendezvous again with the fleet oiler KASKASKIA 300 miles northeast of Fiji Island. Out on the open ocean, 30 April, CV-2 conducted 5 in. gun practice. Also, on that day, VS-2 lost an SBD in a ditching, after a wave-off, while attempting to land, during one of many training flights during this period.

US Naval Intelligence concluded that the Japanese were about to make a move into Solomon Islands, the Coral Sea, Port Moresby, and soon after Australia. US Navy Fleet Admiral Nimitz, in an attempt to repel this move by Japan, organized the arrival of both TF-11 and 17, centered around the aircraft carriers LEX-INGTON and USS YORKTOWN CV-5, to the Coral Sea. He also ordered TF-16, centered around the carriers USS ENTERPRISE CV-6 and USS HORNET CV-8, to arrive as soon as they returned from their assault upon Japan, flying off General Doolittle's B-25 bombers from HORNET. The SARATOGA was still under repair at Puget Sound, USS RANGER CV-4 was delivering US Army P-40 fighters across the Atlantic Ocean, to the Gold Coast of western Africa and USS WASP CV-7 was making her second run of British Spitfire fighters to Malta in the Mediterranean Sea. That left only four carriers to hold back the Japanese advance across the Pacific.

About 0630, TF-11 and 17, as well as the Southwest Pacific Forces, formerly known as the ANZAC Forces, rendezvoused at the predesignated *"Point Buttercup,"* 260 miles northwest of New Caledonia, the two task forces then became one under the designation TF-17. The warships in TF-17 then consisted of the fleet carriers YORKTOWN, LEXINGTON, cruisers CHESTER, CHICAGO, NEW ORLEANS, PORTLAND, ASTORIA, MINNEAPOLIS, 13 destroyers, and 2 fleet oilers. At this time, the task force was to refuel from the two fleet oilers, but this was delayed by LEXINGTON until the next day and beyond.

By 3 May, Admiral Fletcher, aboard YORKTOWN, and TF-17 commander, headed north to scout for Japanese forces, while Admiral Fitch aboard LEXINGTON, with difficulty, continued to refuel. Also, on this day, a Japanese invasion force made an unopposed landing at Tulagi, just to the north of Guadalcanal. As the YORKTOWN group headed north, the LEXINGTON group steamed to the west after fueling was completed in the early afternoon. Late in the afternoon Admiral Fletcher aboard YORKTOWN received word the the Japanese had taken Tulagi. With that news, YORKTOWN's group increased their speed north to 27 kts.

With LEXINGTON's group still traveling west in the Coral Sea, the YORKTOWN group reached a point 100 miles south of Guadalcanal, and just before dawn, at 0630 hrs., launched an air strike on Tulagi. They attacked the Japanese warships in the harbor first with dive bombers, sinking a destroyer and two minesweepers. Next, the torpedo planes swooped in, but only managed to sink another minesweeper. The American planes then headed back south to YORKTOWN, landing by 0930 hrs., not one plane lost.

In the meantime, the LEXINGTON group, out in the Coral Sea, was joined, by the oiler USS NEOSHO AO-23 and a destroyer, at 0800 hrs. Later, at 0900 hrs., the LEXINGTON group was also joined by the Australian cruisers HMAS AUSTRALIA and HOBART and a USN destroyer of TF-44. At that time, the then larger LEXINGTON group, headed to the southeast.

Back at the YORKTOWN, she launched a second strike upon Tulagi at 1030 hrs., but her aircraft only managed to damage a small patrol craft, a destroyer and sink two seaplanes, while losing three of their own aircraft. A third strike was launched at 1400 hrs., sinking four landing barges and all aircraft returned to YORKTOWN by 1632 hrs. At that time Admiral Fletcher turned YORKTOWN south and quickly departed the area, heading towards the LEXINGTON group at high speed.

The LEXINGTON and YORKTOWN groups rendezvoused at 0816 hrs., 5 May 1942, about 200 miles south of Rennell Island. Once they meet, the YORKTOWN group refueled from NEOSHO for most of the day, continuing on a southeasterly course. During this time, in the late morning, one of YORKTOWN's patrolling fighters shot down a Japanese four engine seaplane. After fueling was completed at 1900 hrs., the entire task force turned around and headed northwest at 1930 hrs. in anticipation of Japanese forces advancing from Rabaul, towards Port Moresby, on the south side of New Guinea.

The morning of 6 May 1942 was much the same as the day before. The combined Allied fleet continued to steam in a northwesterly direction, but then proceeded to steam in a very large 80 mile oval, as a holding position in the Coral Sea to the south of Rennell Island. About half way through this course, while heading southeast, at 1100 hrs., another four engined seaplane was spotted, just as it spotted the Allied fleet and then made good it's escape north. Admiral Fletcher, in command of the Allied fleet, new his warships had been spotted and so, at 1755 hrs., the oiler NEOSHO and a destroyer were detached and ordered to steam south to a refueling rendezvous point. The rest of the Allied fleet then steamed at high speed to the northwest in an effort to depart the area where they had been sighted by the Japanese seaplane. They would do so throughout the night.

Just before dawn on 7 May 1942, at 0625 hrs., the American fleet turned due north, but TF-44 Allied Support Group, cruisers AUSTRALIA, CHICAGO and

HOBART, continued on the original course as a diversion and to attack the Japanese invasion force reported heading for Port Moresby. At 0810 hrs., TF-44 encountered an IJN float plane that shadowed them for almost an hour.

Meanwhile, as TF-17 headed north, at 0815 hrs., 7 May 1942, USN carrier reconnaissance spotted, what they thought was the main IJN strike force, just to the north of their position by 175 miles. By 0926 hrs., LEXINGTON launched her strike aircraft, followed about 30min. later by YORKTOWN. LEXINGTON's aircraft spotted the IJN light carrier HIJMS SHOHO and her cruiser escort at about 1100 hrs. and began to attack. At first, no hits were scored, but a near-mis blew five aircraft off the flight deck. At 1110 hrs., SBD dive bombers attacked, TBD torpedo bombers attacked at 1117 hrs., and YORKTOWN's aircraft joined in by 1125 hrs. SHOHO was overwhelmed by at least 93 aircraft, which no warship at that date could withstand. It was estimated that the IJN light carrier was hit by 7 torpedoes and 13 1000 lb. bombs, going dead in the water at 1131 hrs. and sinking by 1136 hrs. at 10°29'S by 152°55'E. This first American sinking of an IJN carrier was a resounding success, with the famous report from one of LEXINGTON's aircraft, "*Scratch one Flattop.*" Both LEXINGTON and YORKTOWN recovered all but 3 aircraft by 1338 hrs. and were ready to launch a second strike by 1450 hrs. to go after the escorting IJN cruisers, but this was called off.

At 1630 hrs., the Japanese fleet carriers, HIJMS SHOKAKU and ZUIKAKU, then unlocated by the Americans, launched 12 dive and 15 torpedo bombers to search for the US Navy carriers. They were unsuccessful, and while returning to their carriers, were discovered by fighters from the American carriers. In the ensuing "Dog Fight," the Japanese lost 9 aircraft, while the Americans lost 2 fighters. As this battle ended, with the oncoming darkness, about 1900 hrs., confusion set in for the Japanese. Many of their aircraft mistook the American carriers as their own, and attempted to land on them! The Americans, quickly realizing the situation, lured them in and managed to shot down 1, possibly 2 aircraft before the Japanese realized their mistake and headed off in the direction of their own carriers. Because of this action, both fleets then knew approximately where the other was. Ironically, they both contemplated sending combined cruiser/destroyer forces against each other, but both decided that it was more advantageous to keep those escorts with their respective fleets. By 2200 hrs., the IJN fleet was headed north, while the USN fleet was headed to the southeast.

As dawn approached on 8 May 1942, LEXINGTON was given the task of searching for the Japanese carriers, launching her aircraft at 0625 hrs. By 0815 hrs., the IJN fleet carriers SHOKAKU and ZUIKAKU were located, both YORKTOWN and LEXINGTON launching their attack aircraft by 0915 hrs. YORKTOWN aircraft started their attack upon SHOKAKU at 1054 hrs., scoring only two bomb hits, one on the flight deck all the way forward and another aft. SHOKAKU could only recover aircraft after that hit. All USN torpedo bombers missed their target. YORKTOWN lost 4 SBDs in this attack. LEXINGTON's aircraft did not fare as well. They had great difficulty with the overcast, some becoming lost in the cloud cover and having to return. About half of her attack aircraft reached the Japanese fleet carriers, attacking at 1140 hrs., scoring only one bomb hit and no torpedo hits on SHOKAKU.

Unknown to the American's, the Japanese had launched a strike of 90 aircraft, from both SHOKAKU and ZUIKAKU, at 0825 hrs. They were headed in the direction of the American carriers, when they spotted USN search planes returning to their carriers and followed them to their targets. The Japanese attack upon both YORKTOWN and LEXINGTON began at 1113 hrs., with torpedo bombers coming in low, from both sides of the bow. YORKTOWN had 8 torpedoes launched at her, but wild maneuvering caused all to mis, but not for LEXINGTON, who could not maneuver nearly as well, was hit by at least 2 of 11 torpedoes launched at her, both to port, one in the bow and the other abreast the island, at 1121 hrs.

No sooner than the Japanese torpedo bomber attack ended, when, at 1124 hrs., the Japanese dive bombers dropped into their attack upon YORKTOWN and LEXINGTON. First YORKTOWN was attacked by numerous dive bombers, with several damaging dear misses, but one managed to hit her flight deck next to the island structure. The bomb exploded on her fourth deck, but the fires were put out quickly and the carrier could still maintain flight operations. LEXINGTON, trying to maneuver, was hit by two bombs, with five damaging near misses. One bomb hit on the edge of the flight deck, port forward, and the other on the funnel. The damage to the flight deck was quickly repaired, but one of the 5 in.

AA guns and it's crew, as well as half the crew to the adjacent mount were killed. The damage from the bomb that hit the funnel was extensive. It hit the port top edge of the funnel, exploding on the .50 cal. machine gun platform, with fragments penetrating to the starboard side, as well as to the back of the bridge and the after directors fitted to the funnel. Many crew in all positions mentioned were killed by this hit. One of the five near misses, near the port forward hinged 20mm platform, severely damaging that platform, putting all those mounts out of action and killing several crew.

By 1132 hrs., the last Japanese aircraft left the area, thus ending the fight. Damage control teams aboard both carriers went about their work to quickly make temporary repairs. LEXINGTON had a 7° list to port, but by 1230 hrs., this was soon corrected by counterflooding hull spaces to starboard. The ship was down by the bow by about 6 ft., but was able to make 25 kts., even though three boilers were put out of action by the port midships torpedo hit. During all this, CV-2 was able to recover her aircraft. It seemed as if the carrier was battle ready again, but in a famous quote from the damage control office, "I would suggest that if you take any more torpedo hits, you take them on the starboard side."

Then suddenly at 1247 hrs. LEXINGTON was shaken by a tremendous explosion, caused by the ignition of gasoline vapors below. Fire raged out of control, with another massive explosion occurring at 1445 hrs., and one more at 1530 hrs. At 1558 hrs. the captain, fearing for the safety of men working below, ordered all hands to the flight deck. Escorting destroyers were alongside, trying to aid the stricken warship. At 1707 hrs., "*Abandon Ship!*", was ordered and by 1727 hrs., just as the last men were leaving the carrier, yet one more massive explosion occurred. LEXINGTON blazed on, into the night, flames shooting hundreds of feet into the air. Destroyer PHELPS closed to 1500 yards and fired two torpedoes into her hull to port and another to starboard, and with one last heavy explosion, the gallant LEXINGTON sank at 1956 hrs. on an even keel, at 15°12'S., by 155°27'E. A total of 2735 crew were rescued by escorting destroyers and cruisers.

LEXINGTON received two battle stars for her World War II service.

The LEXINGTON in the #3 dry dock at the Boston Navy Yard, 12 January 1928, soon after commissioning.

The images on the opposite page are from another sequence of photos taken while CV-2 was in the #3 Drydock at the Boston Navy Yard during the early 1930s. All of the images on these two pages give one an idea as to the massive size of LEXINGTON and her sistership SARATOGA.

One of the SBC-4's (BuNo 1273) from VS-2 suffers a landing accident on USS LEXINGTON (CV-2) on 18 January 1940. This image also offers one of very few views of CV-2's superstructure during that time period.

*USN Photo, Capt Adair Collection, courtesy of Lee Hastings*

The photograph on this page is of the port-side LEXINGTON's bridge, taken 10 December 1933.

USS LEXINGTON CV-2, pierside at the Puget Sound Navy Yard, 26 March 1937, showing the modification to her forward flight deck. This was widened to aid in the launching, landing and deck stowage of her aircraft.

Another view of LEXINGTON at the Puget Sound Navy Yard, this of the entire starboard side of the island structure, showing the bridge and the funnel. This image was taken on the same date as the image on the previous page. Note the boat stowage at the base of the funnel.

A fantastic aerial view of LEXINGTON, 28 April 1938. She was returning to the US West Coast after a fleet exercise out in the Pacific Ocean, seen here, steaming at about 10 knots. The overall painting of the warship at that time was Standard Navy Gray, her flight deck stained Mahogany, with bright yellow deck markings. The open bridge decks would have been covered with a reddish-brown linoleum and any other steel decks on the ship painted Deck Gray, which was a dark gray.

This starboard side aerial view of LEXINGTON, was also taken on 28 April 1938. The openings on the hullside, just below the flight deck were for boat stowage, with her larger 36, 40 and 50 foot long ship's boats seen in place at the taking of this photograph. The smaller guns fitted on the foremost and aftmost flight deck edge platforms, as well as the upper funnel platforms were water cooled .50 cal. anti-aircraft machine guns. The larger guns fitted on the semi-circular flight deck edge platforms were her 5 in./25 cal. heavy anti-aircraft mounts.

# U. S. S. LEXINGTON CV-2

Booklet of General Plans
As Fitted
May 7th., 1941
Puget Sound Navy Yard


Bureau of Ships
**UNITED STATES NAVY**

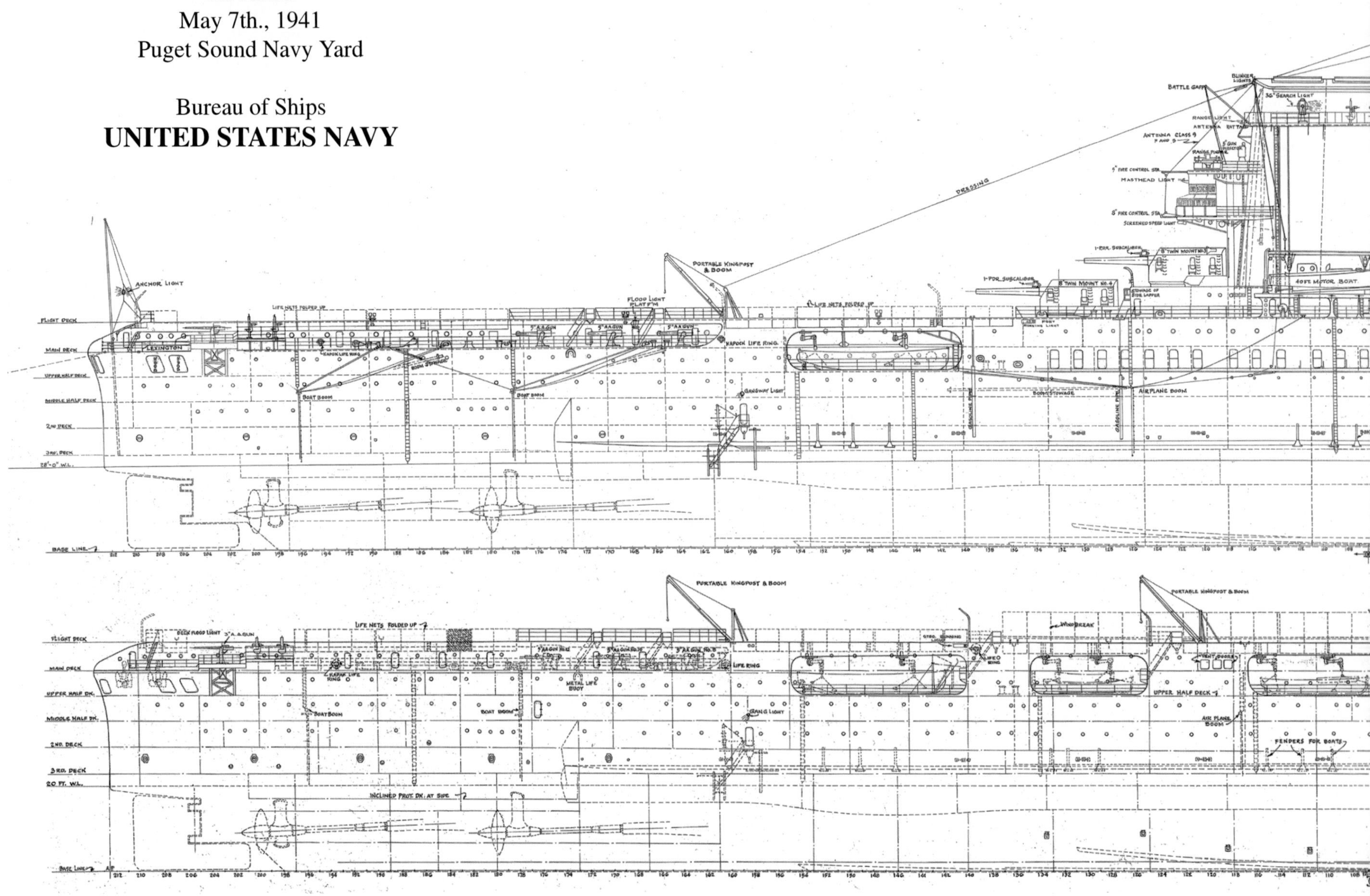

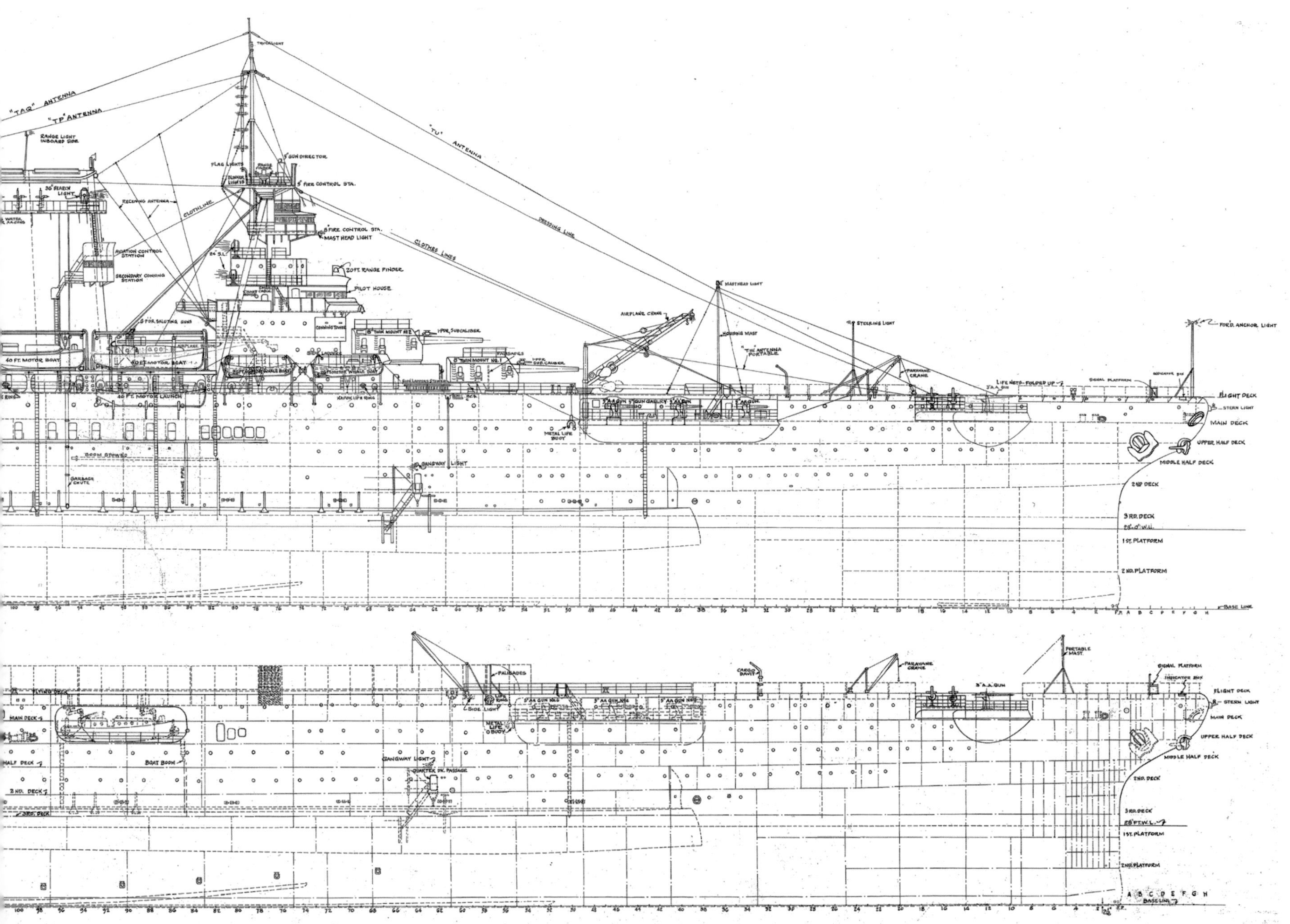
"TAQ" ANTENNA
"TP" ANTENNA
"TU" ANTENNA
RANGE LIGHT INBOARD GUN
36" SEARCH LIGHT
RECEIVING ANTENNA
CLOTH LINE
AVIATION CONTROL STATION
SECONDARY CONNING STATION
FLAG LIGHTS
TRUCK LIGHT
5" GUN DIRECTOR
5" FIRE CONTROL STA.
8" FIRE CONTROL STA.
MAST HEAD LIGHT
DRESSING LINE
CLOTHES LINES
36" S.L.
PILOT HOUSE
20 FT. RANGE FINDER
CONNING TOWER
3" FOR SALUTING GUNS
TWIN MOUNT NO. 2
1 PDR. SUBCALIBER
TWIN MOUNT NO. 1
1 PDR. SUB-CALIBER
MASTHEAD LIGHT
AIRPLANE CRANE
HOUSING MAST
TU ANTENNA PORTABLE
STEERING LIGHT
PARAVANE CRANE
FOR'D ANCHOR LIGHT
LIFE NETS - FOLDED UP
3" A.A. GUN
SIGNAL PLATFORM
5" GUN GALLERY
5" A.A. GUN
40 FT. MOTOR BOAT
50 FT. MOTOR BOAT
40 FT. MOTOR LAUNCH
METAL LIFE BUOY
FLIGHT DECK
STERN LIGHT
MAIN DECK
UPPER HALF DECK
MIDDLE HALF DECK
BOOM STOWED
GANGWAY LIGHT
GARBAGE CHUTE
2ND DECK
3RD DECK
28'-6" F.W.L.
1ST PLATFORM
2ND PLATFORM
BASE LINE
PALISADES
CARGO DAVIT
PARAVANE CRANE
PORTABLE MAST
SIGNAL PLATFORM
INDICATOR BOX
3" A.A. GUN
FLYING DECK
SHOE LIGHT
3" A.A. GUN NO. 3
3" A.A. GUN NO. 2
3" A.A. GUN NO. 1
METAL LIFE BUOY
MAIN DECK
HALF DECK
BOAT BOOM
GANGWAY LIGHT
QUARTER DK. PASSAGE
2ND DECK
3RD DECK
28 FT. W.L.
1ST PLATFORM
2ND PLATFORM
BASE LINE

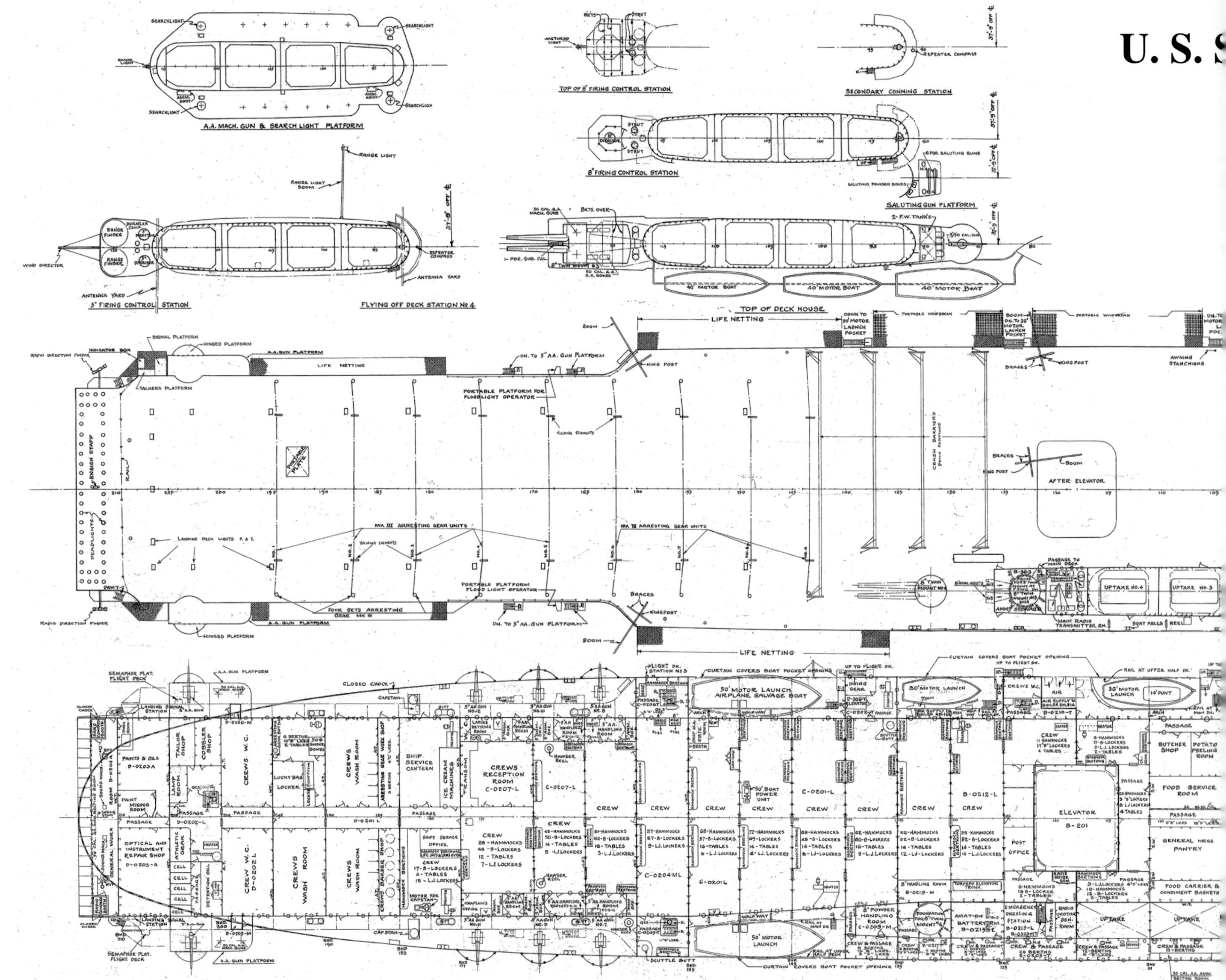

A.A. MACH. GUN & SEARCH LIGHT PLATFORM
SEARCHLIGHT
RANGE LIGHT
TOP OF 8" FIRING CONTROL STATION
8" FIRING CONTROL STATION
SECONDARY CONNING STATION
SALUTING GUN PLATFORM
REPEATER COMPASS
5" FIRING CONTROL STATION
FLYING OFF DECK STATION No 4
RANGE FINDER
WIND DIRECTOR
ANTENNA YARD
40' MOTOR BOAT
TOP OF DECK HOUSE
LIFE NETTING
AFTER ELEVATOR
A.A. GUN PLATFORM
KING POST
PORTABLE PLATFORM FOR FLOODLIGHT OPERATOR
MK III ARRESTING GEAR UNITS
MK IV ARRESTING GEAR UNITS
FOUR SETS ARRESTING GEAR MK III
LANDING DECK LIGHTS P. & S.
RADIO DIRECTION FINDER
HINGED PLATFORM
DN. TO 5" AA. GUN PLATFORM
UPTAKE NO. 4
UPTAKE NO. 3
MAIN RADIO TRANSMITTER RM.
BOAT FALLS REEL
SEMAPHOR PLAT. FLIGHT DECK
PAINTS & OILS
TAILOR SHOP
COBBLER SHOP
CREWS W.C.
CREWS WASH ROOM
SHIP SERVICE CANTEEN
ICE CREAM MACHINES
CREWS RECEPTION ROOM
50' MOTOR LAUNCH AIRPLANE SALVAGE BOAT
30' MOTOR LAUNCH
BUTCHER SHOP
POTATO PEELING ROOM
FOOD SERVICE ROOM
ELEVATOR B-201
GENERAL MESS PANTRY
POST OFFICE
OPTICAL AND INSTRUMENT REPAIR SHOP
PAINT MIXING ROOM
LAMP ROOM
ATHLETIC GEAR
CREW W.C.
CREWS WASH ROOM
BARBER SHOP
SHIPS SERVICE OFFICE
CREW
HAMMOCKS
LOCKERS
TABLES
50' MOTOR LAUNCH
AVIATION BATTERY RM.
EMERGENCY DRESSING STATION
RADIO SEN. ROOM
FOOD CARRIER & CONDIMENT BASKETS
SCUTTLE BUTT
CURTAIN COVERS BOAT POCKET OPENING
LIFE NETTING
UPTAKE
DETENTION CELL
CELL

# LEXINGTON CV-2

Booklet of General Plans

As Fitted

May 7th., 1941

Puget Sound Navy Yard

Bureau of Ships

UNITED STATES NAVY

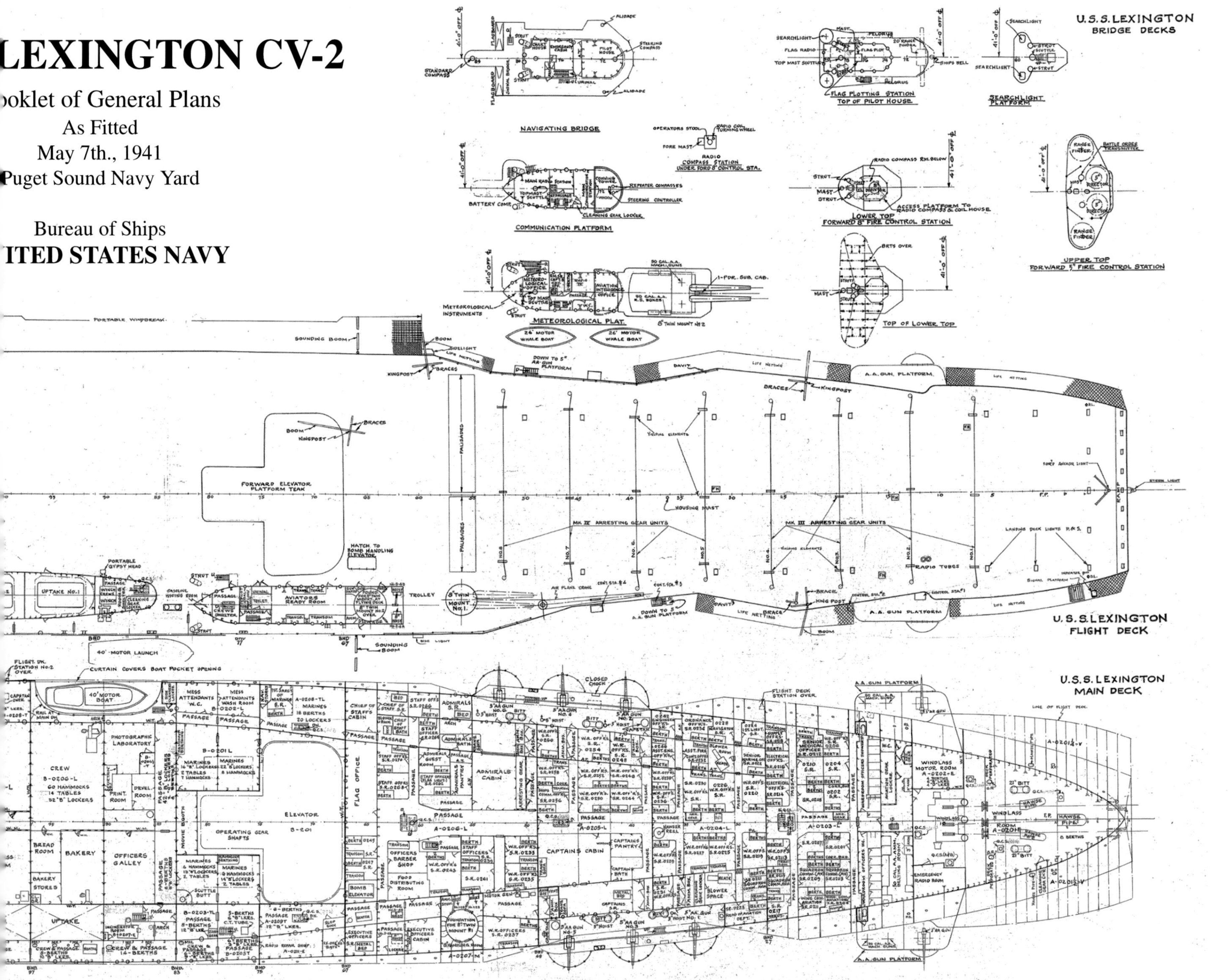

# ALTERATIONS

| NAVY YARD | DATE | DESCRIPTION | APP'RD |
|---|---|---|---|
| N.Y.P.S | 3-26-36 | RETRACED AND CORRECTED TO 10-15-35 AND REVISED TO CONFORM WITH BU. C.&R. LETTER NO. S1-3-(4)(R1) EN7/A2-11-(271) OF 24 JAN. 1934, AND BU. C.&R. LETTER S1-3-(4)(DC) EN7/A2-11 OF 20 MARCH 1935. | R.T.C. 9/20/36 |
| N.Y.P.S | 4/30/37 | SEARCHLIGHT PLATFORM ON STACK DELETED. NEW BARRIERS, ARRESTING GEAR & DECK EXTENSION ADDED TO FLIGHT DECK. COMPT. A-622½-T ADDED IN HOLD & INNER BOTTOM. MISCELLANEOUS COMPT. ASSIGNMENTS AND NUMBERS MODIFIED AND OTHER MINOR CHANGES. CORRECTED TO 4/6/37. | R.T.C. 4/30/37 |
| P.S.N.Y. | 1/31/39 | TOP OF PILOT HOUSE: FLAG PLOT EXTENDED. FLIGHT DK: 40'M.B. CHANGED TO 40' M.L. MAIN DK: HATCH FR.176 R RELOCATED TO STBD. RADIO STOREROOM CHANGED TO MOVIE BOOTH, FR.82 L. UPPER HALF DK: 1ST-LIEUT. OFF. CHANGED TO ASSEMBLY & REPAIR OFF., VF-2 SQUADRON OFF. FR.179 R ADDED. MIDDLE HALF DK: 1ST LIEUT., NAVIGATOR, & DAMAGE CONTROL OFFICES RELOCATED. 2ND & HANGER DK: 2-30' S.W.B. ADDED, D-306-T CHANGED TO D-336-T, D-201-B CHANGED TO D-201-L. 3RD. DK: PORTABLE BAT. CHARGING STATION CHANGED TO ELECT. STOREROOM, FR.132 L. 1ST & 2ND. PLATF.: D-20-F CHANGED TO DIESEL OIL. THE FOLLOWING VOID COMPTS. CHANGED TO FRESH WATER BALLAST TANKS: A-22-W, A-26-W, A-38-W, A-40-W, A-52-W, A-64-W, A-66-W, A-70-W, A-72-W, A-78-W, A-454-W, A-456-W, A-458-W, A-460-W, AND A-462-W. AUTH. BU.C.&R LETTER CV2/S29-8 (C.&DC) OF 7 JAN. 1939. A-622½-T CHANGED TO A-914½-T & OTHER MISC'L MINOR CHANGES. | R.T.C. 1/31/39 |
| P.S.N.Y. | 5-7-41 | 2ND. PLATFORM: CHANGED A-510 M TO 5/50 CAL. A.A. AMM. " D-515A TO D-515M ORDNANCE STORES. CHANGED D-512 A TO 3" A.A. AMM., D-512 M. 3RD. DECK: DELETED TRANSOMS IN C.P.O. QUARTERS. A-308 L, A-309 L AND A-310 L. CHANGED D-343 A TO G.S.K. STORES. CHANGED D-347 A TO SHIPS SERV. STORES. 2ND. DECK: DELETED CAPSTAN ROOM. MIDDLE HALF DECK: CHANGED A-116-TL TO CREW. CHANGED D-110-A TO LAUNDRY ISSUE ROOM. UPPER HALF DECK: RELOCATED W.M. BHD. IN A-0103-A. MAIN DECK: ADDED 3" A.A. GUN ON GUN PLATFORMS FORE AND AFT. EXTENDED FORWARD GUN PLATFORM. FLIGHT DECK: RELOCATED CRASH BARRIERS. ADDED LIFE NETTINGS P.& S. SIDES. BRIDGES: CHANGED TOP OF DECK HOUSE & SALUTING GUN PLATFORM. ADDED 50 CAL. A.A. MACHINE GUNS ON TURRET TOP, AND OTHER MISCELLANEOUS MINOR CHANGES. | R.J.C. 5/7/41 |

# COMPLEMENT

| | |
|---|---|
| COMMANDING OFFICER | 1 |
| WARDROOM OFFICERS | 42 |
| JUNIOR OFFICERS (ENSIGN) | 21 |
| WARRANT OFFICERS | 18 |
| TOTAL SHIPS OFFICERS | 82 |
| AVIATION OFFICERS: | |
| WARDROOM OFFICERS | 94 |
| JUNIOR OFFICERS | 21 |
| TOTAL AVIATION OFFICERS | 115 |
| TOTAL OFFICERS | 197 |

# CREW

| SERVICE BRANCH | SHIP FORCE | | AVIATION | | AIRCRAFT SQUADRONS AND FLAG | | SUMMARY OF C.P.O. & CREW | |
|---|---|---|---|---|---|---|---|---|
| | C.P.O. | CREW | C.P.O. | CREW | C.P.O. | CREW | C.P.O. | CREW |
| SEAMAN | 11 | 467 | 1 | 134 | 3 | 194 | 15 | 873 |
| ARTIFICER | 14 | 182 | 1 | | 12 | 40 | 25 | 211 |
| ENGINEER | 24 | 400 | – | | 1 | 1 | 25 | 440 |
| COMMISSARY | 2 | 26 | – | | | 4 | 2 | 33 |
| SPECIAL | 11 | 85 | 1 | 2 | 8 | 26 | 19 | 112 |
| MESSMEN | – | 53 | | | | 43 | | 92 |
| AVIATION | 11 | 46 | | | 27 | 167 | 38 | 208 |
| MARINE | 2 | 83 | | | | | 2 | 83 |
| TOTAL | | | | | | | 126 | 2047 |

# SMALL BOATS

| NO. | SIZE | MEN EACH | MEN TOTAL |
|---|---|---|---|
| 4 | 50 FT. MOTOR LAUNCHES | 190 | 760 |
| 5 | 40 FT. MOTOR BOATS | 37 | 185 |
| 1 | 40 FT. MOTOR LAUNCHES | 90 | 90 |
| 2 | 36 FT. MOTOR LAUNCH | 70 | 140 |
| 2 | 30 FT. MOTOR LAUNCH | 40 | 80 |
| 2 | 26 FT. MOTOR WHALE BOATS | 22 | 44 |
| 2 | 14 FT. PUNTS | | |
| | TOTAL CAPACITY | | 1299 |

The two documents on this page were part of the Booklet of General Plans from LEXINGTON's May 1941 refit. The leftmost document list the alterations made during that and past refittings, while the document to the right list the ship's personnel and her complement of boats.

This photograph of USS LEXINGTON CV-2 was taken as she entered the Hunter's Point Navy Yard, part of the Mare Island Navy Yard Complex. Many sources cite this image as CV-2 departing San Diego's North Island Naval Air Station on 14 October 1941. It has has been believed that her overall painting was in Ms. 1, consisting of Dark Gray (5-D), which was chalked very badly from weathering. She is also painted with a Ms. 5 False Bow Wave. Her flight deck is dark, which means she could have had that stained with the blue flight deck stain, which was not applied until late October 1941. The carrier was also repainted at that time in Sea Blue (5-S). Because of the evidence in view, there is a chance that this image is actually from sometime in November 1941, giving enough time for the amount of weathering seen here to have taken place. The images on the back cover were taken during her time at Hunter's Point in October 1941. Note the .50 cal. machine guns fitted atop her 8 in. turrets, as well as the elevator wells painted white. This photograph is slightly blurred from the motion of the aircraft that took the image.

One more photograph, newly discovered, of LEXING-
TON at Hunter's Point, loading a Grumman J2F amphibi-
ous aircraft, known as the "DUCK," aboard, during
October 1941. Note how clean the ship's paint work
appears. See the image on the back cover from a color
motion picture film, shot at virtually the same moment
this still photo was taken. Splinter mattresses were also
hung on the railing atop the 8 in. gun mounts, around the
.50 cal. machine guns.

Both of the photographs on this page were taken while LEXINGTON was transporting Marine Corps Vindicator Dive Bombers to Midway Island. She and her escort departed Pearl Harbor on 5 December 1941. In the upper image, CV-2 is with one of her cruiser escorts, USS CHICAGO CA-29.

In this image LEXINGTON has come to almost a complete stop about 450 miles to the east of Midway Island on 7 December 1941, with the news of the attack upon Pearl Harbor. On both this photograph and the one above, a US Navy censor has blotted out the radar antenna on CV-2 and CA-29. Both images were photographed from the cruiser USS PORTLAND CA-33.

The photographs on this page all date from approximately the same time period, early 1942. All of the aircraft in view were painted with the red and white stripped tails, which was started in late December 1941 and was stopped by May 1942. The lower images are still photos pulled from a motion picture film, exact date unknown.

This image was copied out of an old US Navy magazine. The warship in the image was not listed, but one can determine that this is indeed LEXINGTON by the anti-aircraft gun platform fitted high on her funnel. The date of this photograph is also from the late December 1941 thru early May 1942. More than likely, this photo is from the February 1942 period.

USS LEXINGTON CV-2 moored on, or possibly passing by the southwest side of Ford Island, Pearl Harbor Navy Yard, during late March 1942, for what was to be her last refit. Note the boat stowage was previously removed from the base of the funnel, which featured a small funnel cap on the leading uptake vent.

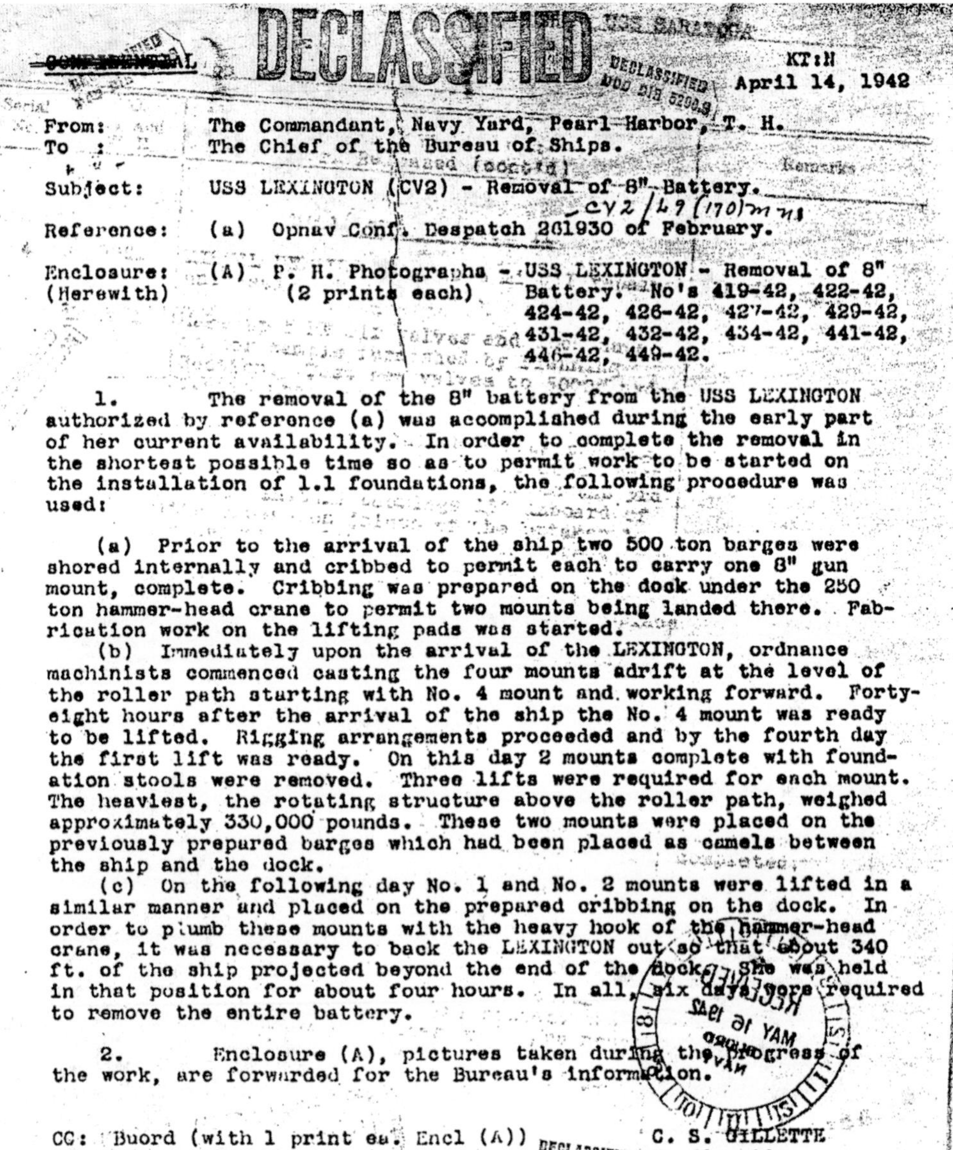

DECLASSIFIED
DOD DIR 5200.9

KT:N

April 14, 1942

From:  The Commandant, Navy Yard, Pearl Harbor, T. H.
To  :  The Chief of the Bureau of Ships.

Subject:  USS LEXINGTON (CV2) - Removal of 8" Battery.

Reference:  (a)  Opnav Conf. Despatch 261930 of February.

Enclosure:  (A)  P. H. Photographs - USS LEXINGTON - Removal of 8"
(Herewith)       (2 prints each)       Battery. No's 419-42, 422-42,
                                       424-42, 426-42, 427-42, 429-42,
                                       431-42, 432-42, 434-42, 441-42,
                                       446-42, 449-42.

1.     The removal of the 8" battery from the USS LEXINGTON authorized by reference (a) was accomplished during the early part of her current availability. In order to complete the removal in the shortest possible time so as to permit work to be started on the installation of 1.1 foundations, the following procedure was used:

(a)  Prior to the arrival of the ship two 500 ton barges were shored internally and cribbed to permit each to carry one 8" gun mount, complete. Cribbing was prepared on the dock under the 250 ton hammer-head crane to permit two mounts being landed there. Fabrication work on the lifting pads was started.

(b)  Immediately upon the arrival of the LEXINGTON, ordnance machinists commenced casting the four mounts adrift at the level of the roller path starting with No. 4 mount and working forward. Forty-eight hours after the arrival of the ship the No. 4 mount was ready to be lifted. Rigging arrangements proceeded and by the fourth day the first lift was ready. On this day 2 mounts complete with foundation stools were removed. Three lifts were required for each mount. The heaviest, the rotating structure above the roller path, weighed approximately 330,000 pounds. These two mounts were placed on the previously prepared barges which had been placed as camels between the ship and the dock.

(c)  On the following day No. 1 and No. 2 mounts were lifted in a similar manner and placed on the prepared cribbing on the dock. In order to plumb these mounts with the heavy hook of the hammer-head crane, it was necessary to back the LEXINGTON out so that about 340 ft. of the ship projected beyond the end of the dock. She was held in that position for about four hours. In all, six days were required to remove the entire battery.

2.     Enclosure (A), pictures taken during the progress of the work, are forwarded for the Bureau's information.

CC: Buord (with 1 print ea. Encl (A))

C. S. GILLETTE
By direction

DECLASSIFIED
DOD DIR 5200.9

051642 40241

U. S. NAVY YARD
PEARL HARBOR, T. H.

DECLASSIFIED
DOD DIR 5200.9

KT:A

C-S74-1/CV/N Y10
Ser.Y-0850

May 1, 1942

From:  The Commandant, Navy Yard, Pearl Harbor, T. H.
To  :  The Chief of the Bureau of Ships.

Subject:  U.S.S. LEXINGTON (CV2) - Installation of 20 MM
          and 1.1 A.A. Guns.

Reference:  (a)  Buships Conf. Despatch 022212 of April, 1942.
            (b)  Comdt. P.H. Conf. ltr C-S74/CV/NY10, Serial
                 Y-0710 of 17 April, 1942.

Enclosure:  (A)  P. H. Photo. #646-42   USS LEXINGTON - 5 - 20
(Herewith)                               MM guns on hinged plat.
2 prints                                 st'bd side. frs 142-152.
each        (B)  P. H. Photo. #647-42   USS LEXINGTON - 3 - 1.1
                                         quad mounts abaft stack
                                         frs. $124\frac{1}{2}$-$133\frac{1}{2}$.
            (C)  P. H. Photo. #648-42   USS LEXINGTON - 2 - 20
                                         MM guns on after end of
                                         stack.
            (D)  P. H. Photo. #649-42   USS LEXINGTON - 20MM gun
                                         at port after corner of
                                         flight deck.
            (E)  P. H. Photo. #650-42   USS LEXINGTON - 20MM guns
                                         at st'bd after corner of
                                         flight deck.
            (F)  P. H. Photo. #651-42   USS LEXINGTON - 6 - 20MM
                                         guns outboard of stack
                                         frs. $98\frac{1}{2}$-$109\frac{1}{2}$. Note clip.
                                         rm. in foreground.
            (G)  P. H. Photo. #652-42   USS LEXINGTON - 4 - 1.1
                                         quad mounts forward of
                                         stack frs. 55-66.
            (H)  P. H. Photo. #653-42   USS LEXINGTON - 4 - 1.1
                                         quad mounts forward of
                                         stack frs. 55-66.
            (J)  P. H. Photo. #654-42   USS LEXINGTON - 4 - 20MM
                                         guns on hinged plat. port
                                         side forward frs. $86\frac{1}{2}$ -
                                         $94\frac{1}{2}$.
            (K)  P. H. Photo. #655-42   USS LEXINGTON - 3 - 20MM
                                         guns on hinged plat. port
                                         side forward frs.127-132.
            (L)  P. H. Plan CV2-78/3-1  20 MM stowage in small
                                        arms mag. compt. A-500-M.
524350      (M)  P. H. Plan CV2-78/3-2  20 MM clip. rms., arrgt.
                                         and details.

DECLASSIFIED

Above are two copies of original USN documents pertaining to CV-2. They provide a partial description of the last refit accomplished on LEXINGTON. In them, a description is given of the work done to remove the 8 in. gun turrets and replace them with the 1.1 in. medium anti-aircraft mounts, as well as the additional 20mm light anti-aircraft mounts that were fitted. In both documents, one can read that photographs of the work accomplished were enclosed. Unfortunately, only the images of the removal of the 8 in. turrets have been located at this time. The photographs of the final fitting of LEXINGTON have yet to be located, if they will ever be found.

Removal of LEXINGTON's 8 inch twin gunned turrets by the huge yard crane at the Pearl Harbor Navy Yard began on 26 March 1942 and finished four days later. Because of the size and capacity of that huge crane, the turrets were lifted in their entirety from the carrier and set onto awaiting barges. The two turrets in this image were mounts #1 and 2. Note that the tops of the gun turrets were painted dark, possibly Deck Blue (20-B) the time of this photograph. The warship itself came into the yard painted in Ms. 12, so that portion of the ship where the turrets were fitted was painted Ocean Gray (5-O).

In this image, 8 inch gun turret #3 was being removed, again by the huge yard crane. Turrets #2 and 3 were fitted with rangefinders, while turrets #1 and 4 were not. The platform attached to the after end of the funnel and just above the 8 inch gun turret being removed was the 8 inch Fire Control Station. Just above that, was the 5 inch AA Fire Control Station platform.

In this, a second photograph of the removal of gun turret #3, is a better view of both the 8 and the 5 inch Fire Control Stations. The use of Splinter Protective Matting was still widespread, as much of this was attached to platform railing and to the sides of smaller enclosed platforms.  The platform high up on the funnel was the .50 cal. AA machine gun gallery, both port and starboard. At the time of the taking of this photograph, repainting of LEXINGTON's camouflage was started, as the dark patch of paint shows on the hull side.

This close-up view of turret #3, as it was lifted into the air from the superstructure of LEXINGTON, shows the multitude of rivets used in the construction of this type of gun mount.

The photograph on the opposite page was of turret #3 being lowered onto the barge placed between the pier and LEXINGTON. Note the de-degaussing cables on the hullside of CV-2. These were massive cables that an electrical current was run through to neutralize the warships magnetic field, in an effort to combat magnetic mines.

Gun turret #3 being lowered onto a barge.

The following page is a close-up image of the face of gun turret #1 being detached from the lifting cradle after it was placed upon the barge.

TURRET 1
ELECTRI

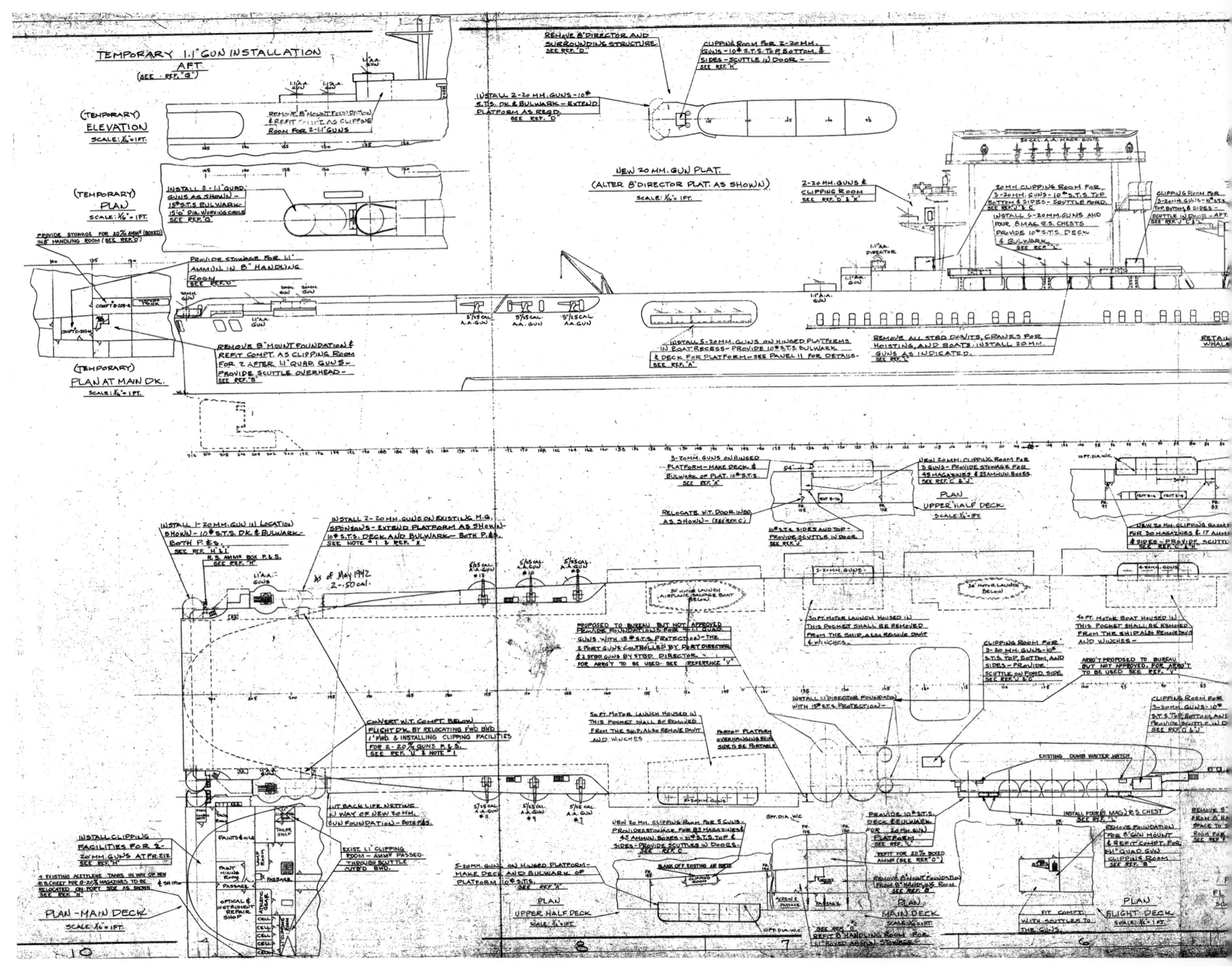

TEMPORARY 1.1" GUN INSTALLATION AFT
(SEE REF. "G")
(TEMPORARY) ELEVATION
SCALE: 1/16" = 1 FT.
(TEMPORARY) PLAN
SCALE: 1/16" = 1 FT.
(TEMPORARY) PLAN AT MAIN DK.
SCALE: 1/16" = 1 FT.
REMOVE B MOUNT FOUNDATION & REFIT COMPT. AS CLIPPING ROOM FOR 2-1.1" GUNS
INSTALL 3-1.1" QUAD. GUNS AS SHOWN - 15# S.T.S. BULWARK 15'-0" DIA. WORKING CIRCLE SEE REF. "G"
PROVIDE STOWAGE FOR 20% AMM'N (BOXED) IN B HANDLING ROOM (SEE REF. "G")
PROVIDE STOWAGE FOR 1.1" AMM'N. IN B HANDLING ROOM SEE REF. "G"
REMOVE B MOUNT FOUNDATION & REFIT COMPT. AS CLIPPING ROOM FOR 2 AFTER 1.1" QUAD GUNS - PROVIDE SCUTTLE OVERHEAD - SEE REF. "B"
REMOVE B DIRECTOR AND SURROUNDING STRUCTURE SEE REF. "D"
CLIPPING ROOM FOR 2-20MM. GUNS - 10# S.T.S. TOP & BOTTOM, & SIDES - SCUTTLE IN DOOR - SEE REF. "K"
INSTALL 2-20 MM. GUNS - 10# S.T.S. DK. & BULWARK - EXTEND PLATFORM AS REQ'D. SEE REF. "O"
NEW 20 MM. GUN PLAT. (ALTER B DIRECTOR PLAT. AS SHOWN)
SCALE: 1/16" = 1 FT.
2-20 M.M. GUNS & CLIPPING ROOM SEE REF. "D" & "K"
50 CAL. A.A. MACH. GUN
20MM. CLIPPING ROOM FOR 3-20MM. GUNS - 10# S.T.S. TOP BOTTOM & SIDES - SCUTTLE FORD. SEE REF. "J" & "C"
INSTALL 6-20MM. GUNS AND FOUR 8 MAG. R.S. CHESTS PROVIDE 10# S.T.S. DECK & BULWARK SEE REF. "L"
CLIPPING ROOM FOR 3-20MM. GUNS - 10# S.T.S. TOP, BOTTOM & SIDES - SCUTTLE IN DOOR
1.1" A.A. DIRECTOR
1.1" A.A. GUN
5"/25 CAL. A.A. GUN
INSTALL 5-20MM. GUNS ON HINGED PLATFORMS IN BOAT RECESS - PROVIDE 10# S.T.S. BULWARK & DECK FOR PLATFORM - SEE PANEL II FOR DETAILS - SEE REF. "A"
REMOVE ALL STBD. DAVITS, CRANES FOR HOISTING, AND BOATS. INSTALL 20MM. GUNS AS INDICATED. SEE REF. "L"
RETAIN WHALE
3-20MM. GUNS ON HINGED PLATFORM - MAKE DECK & BULWARK OF PLAT. 10# S.T.S. SEE REF. "A"
NEW 20MM. CLIPPING ROOM FOR 3 GUNS - PROVIDE STOWAGE FOR 45 MAGAZINES & 25 AMMUN. BOXES SEE REF. "C" & "J"
RELOCATE W.T. DOOR INB'D AS SHOWN - (SEE REF. "C")
10# S.T.S. SIDES AND TOP - PROVIDE SCUTTLE IN DOOR SEE REF. "J"
PLAN UPPER HALF DECK
SCALE: 1/16" = 1 FT.
NEW 20 MM. CLIPPING ROOM FOR 20 MAGAZINES & 17 AMMUN. & SIDES - PROVIDE SCUTTLE
INSTALL 1-20MM. GUN IN LOCATION SHOWN - 10# S.T.S. DK. & BULWARK - BOTH P. & S. SEE REF. "H" & "I"
R.S. AMM'N BOX P. & S. SEE REF. "H"
INSTALL 2-20 MM. GUNS ON EXISTING M.G. SPONSONS - EXTEND PLATFORM AS SHOWN 10# S.T.S. DECK AND BULWARK - BOTH P. & S. SEE NOTE #1 & REF. "X"
AS OF MAY 1942 2-.50 CAL.
5"/25 CAL. A.A. GUN #12
5"/25 CAL. A.A. GUN #10
50 MOTOR LAUNCH AIRPLANE SALVAGE BOAT BELOW
30 FT. MOTOR LAUNCH HOUSED IN THIS POCKET SHALL BE REMOVED FROM THE SHIP, ALSO REMOVE DAVIT & WINCHES
40 FT. MOTOR BOAT HOUSED IN THIS POCKET SHALL BE REMOVED FROM THE SHIP ALSO REMOVE DAVIT AND WINCHES
PROPOSED TO BUREAU BUT NOT APPROVED PROVIDE FOUNDATIONS FOR 2 PORT GUNS WITH 15# S.T.S. PROTECTION - THE 2 PORT GUNS CONTROLLED BY PORT DIRECTOR & 2 STBD GUNS BY STBD. DIRECTOR - FOR ARR'G'T TO BE USED - SEE REFERENCE "V"
CLIPPING ROOM FOR 3-20 MM. GUNS - 10# S.T.S. TOP, BOTTOM, AND SIDES - PROVIDE SCUTTLE IN FORD. SIDE SEE REF. "J" & "C"
ARR'G'T PROPOSED TO BUREAU BUT NOT APPROVED. FOR ARR'G'T TO BE USED SEE REF. "V"
INSTALL 1.1" DIRECTOR FOUNDATION WITH 15# S.T.S. PROTECTION -
CLIPPING ROOM FOR 3-20MM. GUNS - 10# S.T.S. TOP, BOTTOM, AND PROVIDE SCUTTLE SEE REF. "J" & "C"
CONVERT W.T. COMPT. BELOW FLIGHT DK. BY RELOCATING FWD BHD 1' FWD. & INSTALLING CLIPPING FACILITIES FOR 2-20MM GUNS P. & S. SEE REF. "V" & NOTE #1
50 FT. MOTOR LAUNCH HOUSED IN THIS POCKET SHALL BE REMOVED FROM THE SHIP, ALSO REMOVE DAVIT AND WINCHES
PART OF PLATFORM OVERHANGING SHIP SIDE TO BE PORTABLE
EXISTING DUMB WAITER HATCH
INSTALL FOUR 8 MAG. R.S. CHEST SEE REF. "L"
CUT BACK LIFE NETTING IN WAY OF NEW 20MM. GUN FOUNDATION - BOTH P. & S.
5"/25 CAL. A.A. GUN #11
5"/25 CAL. A.A. GUN #9
5"/25 CAL. A.A. GUN #7
NEW 20 MM. CLIPPING ROOM FOR 5 GUNS - PROVIDE STOWAGE FOR 85 MAGAZINES 45 AMMUN. BOXES - 10# S.T.S. TOP & SIDES - PROVIDE SCUTTLES IN DOORS
PROVIDE 10# S.T.S. DECK & BULWARK FOR 20MM. GUN PLATFORM SEE REF. "J"
REMOVE B MOUNT FOUNDATION & REFIT COMPT. FOR CLIPPING ROOM SEE REF. "B"
INSTALL CLIPPING FACILITIES FOR 2-20MM. GUNS AT FR. 212 SEE REF. "H"
PAINTS & OILS
TAILOR SHOP
PAINT MIXING ROOM
PASSAGE
OPTICAL & INSTRUMENT REPAIR SHOP
EXIST. 1.1" CLIPPING ROOM - AMM'N PASSED THROUGH SCUTTLE CUT IN D BHD.
5-20MM. GUNS ON HINGED PLATFORM - MAKE DECK AND BULWARK OF PLATFORM 10# S.T.S. SEE REF. "A"
BLANK OFF EXISTING AIR PORTS
CLIPPING ROOM
REMOVE B MOUNT FOUNDATION FROM B HANDLING ROOM SEE REF. "B"
4 EXISTING ACETYLENE TANKS IN WAY OF NEW 8 CHEST FOR 2-20% MAGAZINES TO BE RELOCATED ON PORT SIDE AS SHOWN SEE REF. "N"
PLAN - MAIN DECK
SCALE: 1/16" = 1 FT.
PLAN UPPER HALF DECK
SCALE: 1/16" = 1 FT.
PLAN MAIN DECK
SCALE: 1/16" = 1 FT.
PLAN FLIGHT DECK
SCALE: 1/16" = 1 FT.
FIT COMPT. WITH SCUTTLES TO THE GUNS.
REFIT B HANDLING ROOM FOR 1.1" BOXED AMM'N STOWAGE
10 FT. DIA. W.C.

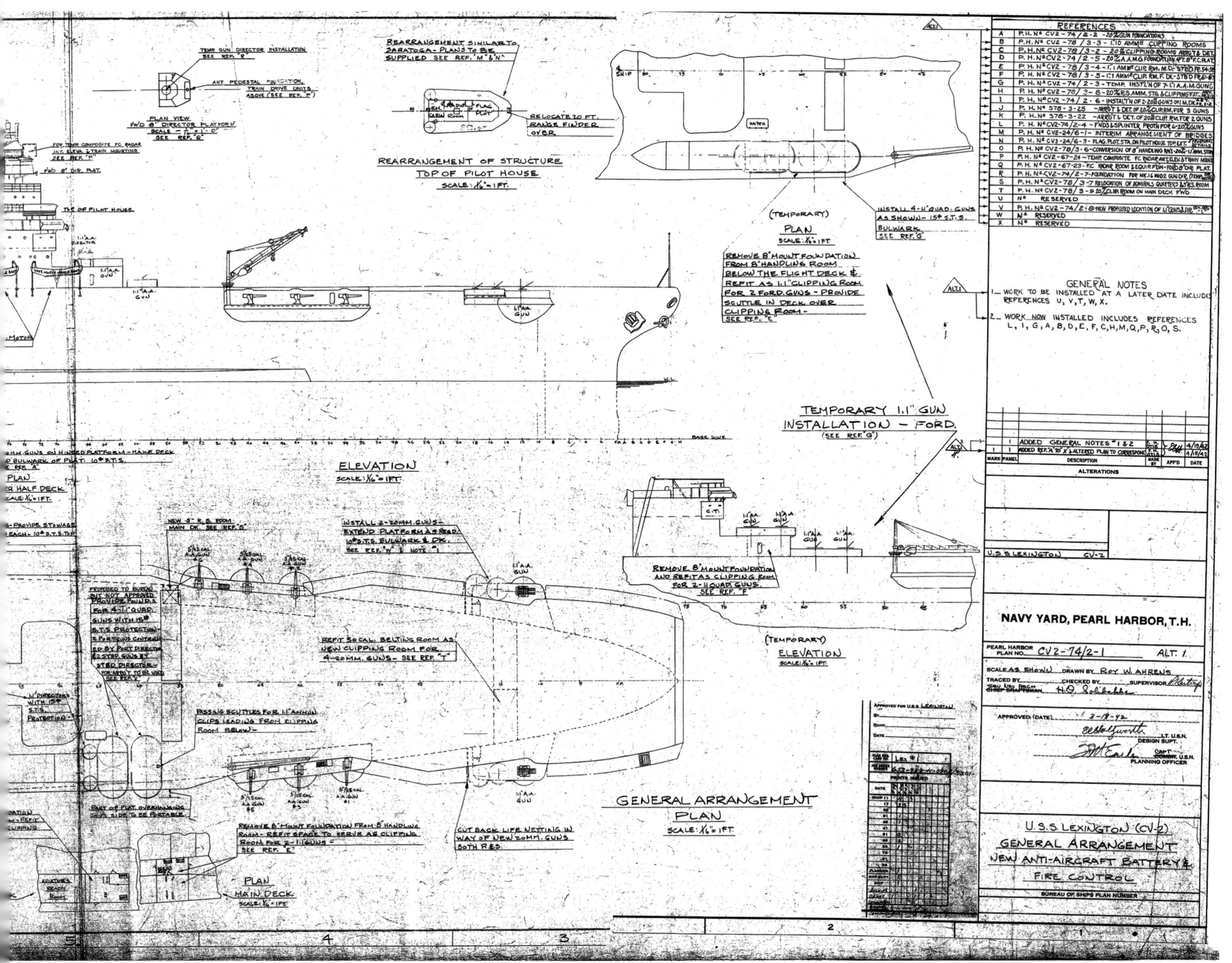

REFERENCES
A  P.H. Nº CV2-74/2-2 - 20% GUN FOUNDATIONS
B  P.H. Nº CV2-78/3-3 - 1.10 AMM'N CLIPPING ROOMS
C  P.H. Nº CV2-74/3-2 - 20% CLIPPING ROOMS ARR'GT & DET.
D  P.H. Nº CV2-74/2-5 - 20% A.A.M.G. FOUNDATION AFT. 8" F.C. PLAT.
E  P.H. Nº CV2-78/3-4 - 1.1 AMM'N CLIP. RM. M.DK. ST'BD. FR. 54-58
F  P.H. Nº CV2-78/3-5 - 1.1 AMM'N CLIP. RM. F. DK. ST'BD. FR. 61-67
G  P.H. Nº CV2-74/2-3 - TEMP. INST'L'N OF 7-1.1 A.A.M. GUNS
H  P.H. Nº CV2-78/3-8 - 20% R.S. AMM. STO. & CLIPPING FF'
I  P.H. Nº CV2-74/2-6 - INSTALL'N OF 2-20% GUNS ON M.DK. FR. 212
J  P.H. Nº 378-3-25 - ARR'GT & DET. OF 20% CLIP.RM. FOR 3 GUNS
K  P.H. Nº 378-3-22 - ARR'GT. DET. OF 20% CLIP. RM. FOR 2 GUNS
L  P.H. Nº CV2-74/2-4 - FND'S & SPLINTER PROT'N FOR 6-20% GUNS
M  P.H. Nº CV2-24/6-1 - INTERIM ARRANGEMENT OF BRIDGES
N  P.H. Nº CV2-24/6-9 - FLAG. PLOT STA. ON PILOT HOUSE TOP EXT.
O  P.H. Nº CV2-78/3-6 - CONVERSION OF 8" HANDLING RM.S. 2-1.1" AMM. STRM
P  P.H. Nº CV2-67-24 - TEMP COMPOSITE FC RADAR ANT. ELEV. & TRAIN MNT.
Q  P.H. Nº CV2-67-23 - FC RADAR ROOM & EQUIP FDN - FOR'D 8" DIR. PLAT.
R  P.H. Nº CV2-74/2-7 - FOUNDATION FOR MK. 16 MOD.2 GUN DIR. (17MP)
S  P.H. Nº CV2-78/3-7 - RELOCATION OF ADMIRALS QUARTERS & R.S. ROOM
T  P.H. Nº CV2-78/3-9 20% CLIP. ROOM ON MAIN DECK FWD
U  Nº  RESERVED
V  P.H. Nº CV2-74/2-8 - NEW PROPOSED LOCATION OF 4" GUNS & DIR
W  Nº  RESERVED
X  Nº  RESERVED


GENERAL NOTES
1 - WORK TO BE INSTALLED AT A LATER DATE INCLUDES REFERENCES U, V, T, W, X.
2 - WORK NOW INSTALLED INCLUDES REFERENCES L, I, G, A, B, D, E, F, C, H, M, Q, P, R, O, S.


TEMP. GUN DIRECTOR INSTALLATION SEE REF. "R"
ANT. PEDESTAL FOUNDATION. TRAIN DRIVE UNITS ABOVE (SEE REF. "P")
PLAN VIEW FW'D 8" DIRECTOR PLATFORM SCALE - 3/8" = 1'-0" SEE REF. "Q"
FOR TEMP. COMPOSITE F.C. RADAR ANT. ELEV'N & TRAIN MOUNTING SEE REF. "P"
FW'D 8" DIR. PLAT.
TOP OF PILOT HOUSE.
1.1" A.A. DIRECTOR
1.1" A.A. GUN
1.1" A.A. GUN
REARRANGEMENT SIMILAR TO SARATOGA - PLANS TO BE SUPPLIED SEE REF. "M" & "N"
FLAG PLOT
CREW ROOM
RELOCATE 20 FT. RANGE FINDER OVER
REARRANGEMENT OF STRUCTURE TOP OF PILOT HOUSE SCALE: 1/16" = 1 FT.
HATCH
INSTALL 4-4" QUAD. GUNS AS SHOWN - 15# S.T.S. BULWARK SEE REF. "G"
PLAN SCALE: 1/16" = 1 FT
REMOVE 8" MOUNT FOUNDATION FROM 8" HANDLING ROOM BELOW THE FLIGHT DECK & REFIT AS 1.1" CLIPPING ROOM FOR 2 FOR'D. GUNS - PROVIDE SCUTTLE IN DECK OVER CLIPPING ROOM - SEE REF. "E"
1.1" A.A. GUN
TEMPORARY 1.1" GUN INSTALLATION - FORD. (SEE REF. "G")
C.T.
1.1" A.A. GUN
1.1" A.A. GUN
ELEVATION SCALE: 1/16" = 1 FT.
BASE LINE
REMOVE 8" MOUNT FOUNDATION AND REFIT AS CLIPPING ROOM FOR 2-1.1 QUAD GUNS. SEE REF. "F"
(TEMPORARY) ELEVATION SCALE: 1/16" = 1 FT.
NEW 5" R.S. ROOM MAIN DK. SEE REF. "S"
INSTALL 2-20MM. GUNS - EXTEND PLATFORM AS REQ'D. 10# S.T.S. BULWARK & DK. SEE REF. "W" & NOTE 1
5"/51 CAL A.A. GUN
5"/51 CAL A.A. GUN #3
5"/51 CAL A.A. GUN #2
1.1" A.A. GUN
PROPOSED TO BUREAU BUT NOT APPROVED PROVIDE FOUND'S FOR 4-1.1" QUAD GUNS WITH 15# S.T.S. PROTECTION
REMOVE 8" MOUNT FOUNDATION AND REFIT AS CLIPPING ROOM FOR 2-1.1 QUAD GUNS. SEE REF. "F"
REFIT 50 CAL BELTING ROOM AS NEW CLIPPING ROOM FOR 4-20MM. GUNS - SEE REF. "T"
PASSING SCUTTLES FOR 1.1" AMMUN. CLIPS LEADING FROM CLIPPING ROOM BELOW.
PART OF PLAT. OVERHANGING SHIP'S SIDE TO BE PORTABLE.
5"/51 CAL A.A. GUN
5"/51 CAL A.A. GUN
REMOVE 8" MOUNT FOUNDATION FROM 8" HANDLING ROOM - REFIT SPACE TO SERVE AS CLIPPING ROOM FOR 2-1.1" GUNS - SEE REF. "E"
CUT BACK LIFE NETTING IN WAY OF NEW 20MM. GUNS BOTH P. & S.
1.1" A.A. GUN
GENERAL ARRANGEMENT PLAN SCALE: 1/16" = 1 FT.
PLAN MAIN DECK SCALE: 1/16" = 1 FT.
AVIATORS READY ROOM


NAVY YARD, PEARL HARBOR, T.H.
PEARL HARBOR PLAN NO. CV2-74/2-1   ALT. 1
SCALE AS SHOWN   DRAWN BY Roy W. Ahrens
TRACED BY
CHECKED BY
SUPERVISOR
CHIEF DRAFTSMAN  H.O. Schlobke
U.S.S. LEXINGTON  CV-2
APPROVED FOR U.S.S. LEXINGTON
BY
RANK
DATE
APPROVED (DATE)  3-18-42
LT. U.S.N. DESIGN SUPT.
CAPT. MEMBER, U.S.N. PLANNING OFFICER
PRINTS ISSUED
U.S.S. LEXINGTON (CV-2)
GENERAL ARRANGEMENT NEW ANTI-AIRCRAFT BATTERY & FIRE CONTROL
BUREAU OF SHIPS PLAN NUMBER


I  ADDED GENERAL NOTES "1 & 2"  4/15/42
ADDED REF. "A" TO "X" & ALTERED PLAN TO CORRESPOND  4/18/42
MARK PANEL  DESCRIPTION  MADE BY  APP'D  DATE
ALTERATIONS

The upper photograph is of the barges with gun turrets #3 and 4 with their lower power mechanisms aboard.

The lower photograph is of LEXINGTON with all 8 inch gun turrets removed, as well as the progression of the change in camouflage painting from Ms. 12 to Ms. 11.  Measure 11 camouflage consisted of painting all vertical surfaces Navy Blue (5-N), as Sea Blue (5-S) had been phased out by that time. Dark patches of Navy Blue are visible on the hullside below the bridge and just aft of the funnel.

# NAVY YARD, PEARL HARBOR, T.H.

PEARL HARBOR PLAN NO. CV2-74/2-1     ALT. 1.

SCALE AS SHOWN     DRAWN BY ROY W. AHRENS
TRACED BY _____     CHECKED BY _____     SUPERVISOR H. Hartrup
SEN. NAV. ARCH. CHIEF DRAFTSMAN     H. O. Solibalke

APPROVED: (DATE) 3-18-42

E. B. Holtzworth ____ LT. U.S.N.
DESIGN SUPT.

J. W. Earle ____ CAPT. COMDR. U.S.N.
PLANNING OFFICER

## U.S.S. LEXINGTON (CV-2)
## GENERAL ARRANGEMENT
## NEW ANTI-AIRCRAFT BATTERY & FIRE CONTROL

BUREAU OF SHIPS PLAN NUMBER

| REFERENCES | |
|---|---|
| A | P.H. Nº CV2-74/2-2 -20 ⁄ₘ GUN FOUNDATIONS |
| B | P.H. Nº CV2-78/3-3 - 1.10 AMMᴺ CLIPPING ROOMS |
| C | P.H. Nº CV2-78/3-2 - 20 ⁄ₘ CLIPPING ROOMS ARR'G'T & DET. |
| D | P.H. Nº CV2-74/2-5 - 20 ⁄ₘ A.A.M.G. FOUNDATION AFT. 8" F.C. PLAT. |
| E | P.H. Nº CV2-78/3-4 - 1".1 AMMᴺ CLIP. RM. M.D. - STB'D FR.54-58 |
| F | P.H. Nº CV2-78/3-5 - 1".1 AMMᴺ CLIP. RM. F. DK. - STB'D FR.61-67 |
| G | P.H. Nº CV2-74/2-3 - TEMP. INSTL'N OF 7-1".1 A.A.M. GUNS |
| H | P.H. Nº CV2-78/3-8 - 20 ⁄ₘ R.S. AMM. STO. & CLIPPING FAC., ABT. FR. 212 |
| I | P.H. Nº CV2-74/2-6 - INSTAL'T'N OF 2-20 ⁄ₘ GUNS ON M.DK. ABT. FR. 212 |
| J | P.H. Nº S78-3-25  - ARR'G'T & DET. OF 20 ⁄ₘ CLIP. RM. FOR 3 GUNS |
| K | P.H. Nº S78-3-22  - ARR'G'T & DET. OF 20 ⁄ₘ CLIP. RM. FOR 2 GUNS |
| L | P.H. Nº CV2-74/2-4 - FND'S & SPLINTER PROT'N FOR 6-20 ⁄ₘ GUNS |
| M | P.H. Nº CV2-24/6-1 - INTERIM ARRANGEMENT OF BRIDGES |
| N | P.H. Nº CV3-24/6-3 - FLAG. PLOT. STA. ON PILOT HOUSE TOP EXT. STRUCTURAL DETAILS |
| O | P.H. Nº CV2-78/3-6 - CONVERSION OF 8" HANDLING RM'S -20 ⁄ₘ-1.1" AMM. STOW |
| P | P.H. Nº CV2-67-24 - TEMP. COMPOSITE FC RADAR ANT. ELEV. & TRAIN MOUNT |
| Q | P.H. Nº CV2-67-23 - F.C. "RADAR ROOM & EQUIP. F'DN - FOR'D 8" DIR. PLAT. |
| R | P.H. Nº CV2-74/2-7 - FOUNDATION FOR MK. 16 MOD 2 GUN DIR. (TEMP. DIR. INSTAL) |
| S | P.H. Nº CV2-78/3-7 RELOCATION OF ADMIRALS QUARTERS & 5" R.S. ROOM |
| T | P.H. Nº CV2-78/3-9 20 ⁄ₘ CLIP. ROOM ON MAIN DECK FWD |
| U | Nº RESERVED |
| V | P.H. Nº CV2-74/2-8 NEW PROPOSED LOCATION OF 1.1" GUNS & DIR. GEN. ARR'G'T "C" |
| W | Nº RESERVED |
| X | Nº RESERVED |

## GENERAL NOTES

1 — WORK TO BE INSTALLED AT A LATER DATE INCLUDES REFERENCES U, V, T, W, X.

2 — WORK NOW INSTALLED INCLUDES REFERENCES L, I, G, A, B, D, E, F, C, H, M, Q, P, R, O, S.

| MARK | PANEL | DESCRIPTION | MADE BY | APP'D | DATE |
|---|---|---|---|---|---|
| 1 | | ADDED GENERAL NOTES #1 & 2 | F.H. OTIS | Elott | 4/15/42 |
| 1 | 1 | ADDED REF. "A" TO "X" & ALTERED PLAN TO CORRESPOND | F.H. OTIS | Elott | 4/15/42 |

Both of the documents on this page were part of the refitting plans from CV-2's final refit, describing the work to be done and that which was to be accomplished at a later date when the warship would become available for yard time. These items are reflected in the drawings dated April 1942, shown on the previous pages.

In this and the two photos on the following page, LEX-INGTON maneuvers wildly in an attempt to avoid torpedoes and bombs from attacking Japanese carrier aircraft. The attack came shortly before 12 noon, 8 May 1942. Japanese "Kate" torpedo bombers and "Val" dive bombers attacked simultaneously, hitting CV-2 with two torpedoes, one midships and one forward on her port side. The dive bombers hit her with one bomb on the flight deck, the other on the island and several damaging near misses. Note the Japanese torpedo bomber just off LEXING-TON's bow.

LEXINGTON maneuvering to avoid Japanese dive bombers, the near misses throwing up huge columns of spray. She was such a heavy and long ship that she was slow to maneuver, especially in the large swells of the Pacific Ocean. For the Japanese torpedo and dive bomber pilots, she was a very large target.

Smoke billowing up from the bomb hit at the base of her funnel, LEXINGTON continues to maneuver to avoid the attacking Japanese carrier aircraft. Note the diving Japanese aircraft just above this text.

Below, a pair of escorting destroyers rushing to aid the then crippled carrier, as she began to slow after taking two torpedo and two bomb hits.

LEXINGTON still able to make at least 25 knots, as USN "Wildcat" fighters fly overhead to protect their "nest." She was smoking heavily from her funnel being hit by a bomb in this attack. The time was shortly after 12 noon, 8 May 1942.

Just landed from their attack upon the SHOKAKU, a TBD taxies on the flight deck as an F4F-3 is about to land on the afternoon of 8 May 1942, at about 1400 hrs. This view is looking aft on the ship's port side, along the bomb damaged 5 in. AA gun gallery.

The crew continue to inspect the damage to LEXING-TON, starboard side, aft. The 5 in. gun mount in the fore ground was put out of action by a bomb hit, visible just beyond that gun. Repair crews work to repair the damage and make the warship battle-ready.

Another view of the port 5 in. gun gallery and it's damage. The two guns in the foreground were still operational, the next was put out of action. The white on the flight deck and gun mount is fire retardant foam.

One more view of the damage to the port forward 5 inch AA gun gallery. Repair crews work to clean up and repair what damage they are able to, in an effort to get the carrier ready for flight operations.

Damage to the port forward 5 inch AA gun gallery. This was caused by a bomb from a Japanese dive bomber that hit near the #4 gun on the after end of that gun gallery. The white spray over many of the items in the photograph is from fire suppression foam, which helps to deprive a fire of the air it needs to burn.

The image on the following page was actually taken aboard the aircraft carrier USS ENTERPRISE CV-6, of one of her .50 cal. machine gun galleries. This was very similar to the gallery near the top of LEXINGTON's funnel, and what it and gun crews would have looked like during a lull in the battle.

LEXINGTON about 1430hrs., 8 May 1942, just after recovering Torpedo Bomber Squadron Two and Fighting Squadron Two. She is down by the bow from a torpedo hit. It is believed that her camouflage painting at that time was Ms. 11, overall Navy Blue (5-N), flight deck stained dark blue and all steel decks Deck Blue (20-B). About the time of the attack upon Pearl Harbor, the linoleum had been removed from all decks. This photograph was taken from the heavy cruiser USS PORTLAND CA-33.

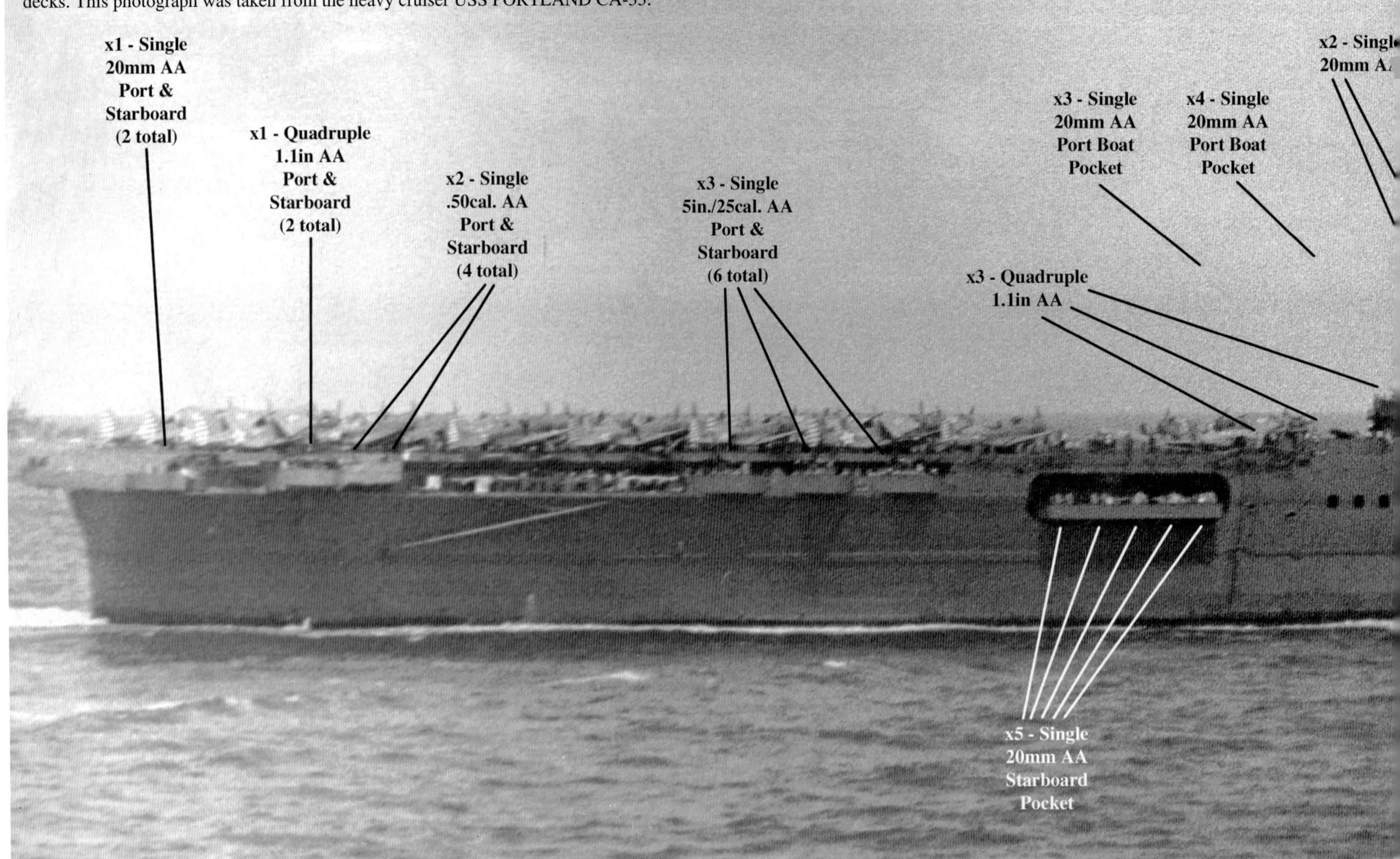

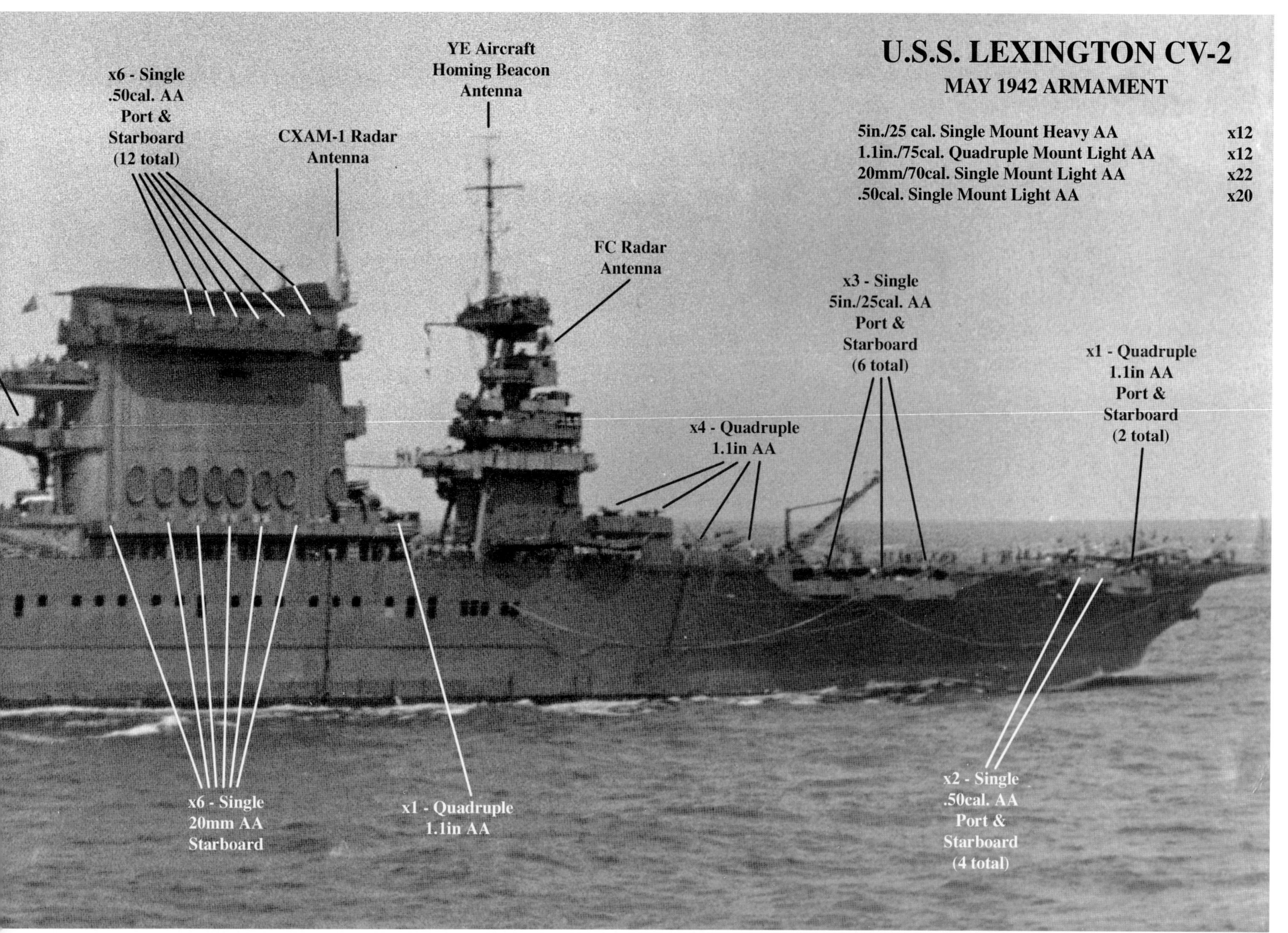

x6 - Single
.50cal. AA
Port &
Starboard
(12 total)

CXAM-1 Radar
Antenna

YE Aircraft
Homing Beacon
Antenna

FC Radar
Antenna

U.S.S. LEXINGTON CV-2
MAY 1942 ARMAMENT

5in./25 cal. Single Mount Heavy AA                    x12
1.1in./75cal. Quadruple Mount Light AA            x12
20mm/70cal. Single Mount Light AA                  x22
.50cal. Single Mount Light AA                             x20

x3 - Single
5in./25cal. AA
Port &
Starboard
(6 total)

x1 - Quadruple
1.1in AA
Port &
Starboard
(2 total)

x4 - Quadruple
1.1in AA

x6 - Single
20mm AA
Starboard

x1 - Quadruple
1.1in AA

x2 - Single
.50cal. AA
Port &
Starboard
(4 total)

The photograph on the left page is of the torpedo hit on LEXINGTON's portside, looking through the damaged flight deck life netting. This hit, midships, on her anti-torpedo bulge, blew some of the hull plating upward, visible in this image.

The photograph on this page was from the 01 level on the island structure about 1630 to 1700 hrs. as the crew muster on the flight deck in preparation to abandon ship. A series of explosions aboard LEXINGTON, mortally crippling her, the captain had little choice, but to execute that order. The heavy cruiser USS MINNEAPOLIS CA-36 passed by forward, seen through the haze of smoke. Note the rectangular shape to the gun tub on the far right edge of this image, as well as the tips of the 1.1 inch AA mount in that tub. The flight deck aircraft crane is in the haze just beyond.

The scene aboard LEXINGTON was grim. In this photo one can see the smoke from the fires below emitting from the flight deck elevator. The fires were not under control in the hangar spaces.

LEXINGTON, viewed from the heavy cruiser MINNEAPOLIS, began to abandon ship, as fires and explosions mortally wounded the once gallant carrier.

This image is a close-up view of LEXINGTON enlarged
from the photograph on the page to the left. In this
enlargement one can see many of the crew on the aft flight
deck, beginning to abandon ship. The destroyer alongside,
starboard, is one from the SIMS class, possibly USS
MORRIS DD-417, attempting to aid in the fire fighting.

LEXINGTON, again viewed from the heavy cruiser MIN-
NEAPOLIS, began to abandon ship, as fires and explosions mortally wounded the once gallant carrier. The
MORRIS was still along side at this point. Crew are drop-
ping down to the water by rope, to be rescued by small
boats from the cruiser and the destroyers.

In this photograph of LEXINGTON, many of the crew are lowering themselves down to the sea by rope. Note the smoke from the hangar fires emitting from the starboard midships 20mm gallery, formally a boat well.

LEXINGTON, in two more views from the heavy cruiser MINNEAPOLIS, in the process of abandoning ship. The MORRIS is still along the starboard side trying to fight fires, while yet another SIMS class destroyer pulls up to CV-2's port stern to assist in the rescue of her crew.

This image is an enlargement from another of the many photographs taken of the abandonment of LEXINGTON. In this image one can see many of the details, such as the degaussing cables, and the supports for the 1.1 in. quadruple, 20mm single and .50 cal. single AA mounts. The crew were being rescued by a 40ft. launch from the cruiser MINNEAPOLIS.

The rescue destroyers backed away from the burning carrier due to the possibility of danger from exploding ordnance. The hangar fire was out of control, as seen by the increase in smoke emitting from the stricken vessel.

LEXINGTON, in another view from the heavy cruiser MINNEAPOLIS, abandoning ship, as fires grow larger. MORRIS was backing away, with either USS ANDERSON DD-411, or HAMMANN DD-412 off the portside of CV-2. One of MINNEAPOLIS's boats approached with a load of survivors, as other LEXINGTON crew lower themselves from the flight deck to the sea.

The motion blur in this photograph is probably not from
the concussion of the explosion that just rocked LEXING-
TON, but more likely just a nervous cameraman trying to
quickly get a shot of that event. It is not certain where the
explosion took place, as it appears that the black smoke
was emitting from the funnel, but it could also have come
from the elevator pit located adjacent to the island, just on
the other side of the funnel.

This distant view of LEXINGTON beginning her death throws, taken by another US Navy heavy cruiser, shows not only CV-2, but two NEW ORLEANS class heavy cruisers, USS CHESTER CA-27 and three out of the four SIMS class destroyers escorting the carrier battle group.

A 40 ft. launch from the MINNEAPOLIS, fully loaded, passed the cameraman in this image, while it was headed towards the destroyer MORRIS to off-load CV-2 survivors onto that warship.

This stern-on view of LEXINGTON, then starting to burn more heavily, shows the destroyer HAMMANN backing away to avoid any unexpected explosions and let the small boats do the close-in rescue work.

The destroyers MORRIS and HAMMANN backing away from the carrier LEXINGTON and the ever increasing fires aboard that stricken warship. Note the rescued crewmen from CV-2 on the stern of MORRIS.

LEXINGTON started to burn heavily at that time, with much more smoke pouring from many openings, as the heavy cruiser MINNEAPOLIS backed away. Note the carrier YORKTOWN and her escort on the horizon.

Top image; Another large explosion rocks the LEXINGTON, as three SIMS class destroyers and a NEW ORLEANS class heavy cruiser head, or back away from the stricken aircraft carrier.

Three Lower Images: This is a series of photographs of a massive detonation of ordnance deep in the hangar deck. The fires that ragged out of control eventually reached the aircraft ordnance, which exploded at about 1727 hrs.

This is the best quality image of the terrific explosion of the aircraft ordnance aboard LEXINGTON that this author was able to find. At first I thought that the dust and hair in the image was just from poor reproduction, but soon realized that these were actually the wood planks from the flight deck that were raining back down to Earth from the massive explosion itself. Note also the splashes in the sea from other objects blown up into the air from that same explosion. Visible in the background is a SIMS class destroyer and on the horizon, the USS YORKTOWN CV-5, headed away from the smoke from LEXINGTON. Also note that a few small boats were still rescuing survivors under the carriers stern.

LEXINGTON, now well on fire and listing, is beginning to sink more rapidly. The fires from the hangar deck were then beginning to spread up to the flight deck. This was due to the exploding aircraft ordnance, fueled by aviation gasoline.

Numerous fires can be seen in this photograph of LEX-
INGTON burning. The explosion of her aircraft ordnance
must have gutted the midships portion of the once mighty
aircraft carrier, as she began to list and sink more rapidly
at this point.

This photograph, snapped just seconds after another series of terrific explosions amidships on LEXINGTON, send up a massive cloud of billowing smoke. Note the debris splashing in the sea all around the stricken carrier. Even though there is a significant amount of motion blur in this image, one of the cruisers 40ft. launches can be seen in the lower left of this image. It must have been waiting to see if there were more survivors to rescue.

# USS LEXINGTON CV-2   AIR GROUPS

**1927**   Scouting Fleet (August)

USS LEXINGTON CV-2 at Quincy Mass.

| | |
|---|---|
| VJ-2S | 1x T3M-1 |
| VT-1S | 16x T3M-2 |
| Utility Unit | 2x UO-1C |

**1928**   Battle Fleet (August)

USS LEXINGTON CV-2

| | |
|---|---|
| VF-3B | 12x FB-5, 1x UO-1 |
| VB-1B | 2x F6C-2, 17x F6C-3, 1x UO-1C |
| VS-3B | No aircraft assigned |
| VT-1B | 9x T3M-2, 10x T4M-1 |
| Utility Unit | 3x O2U-1, 1x T3M-2 |

**1929**   Battle Fleet (April)

USS LEXINGTON CV-2

| | |
|---|---|
| VF-3B | 12x F3B-1, 1x FU-2 |
| VB-1B | 15x F3B-1 |
| VS-3B | 12x O2U-2 |
| VT-1B | 28x T4M-1 |
| Utility Unit | 1x UO-1, 3x O2U-1 |

**1930**   Battle Fleet (June)

USS LEXINGTON CV-2

| | |
|---|---|
| VF-3B | 20x F3B-1 |
| VF-5B | 18x F4B-1 |
| VS-3B | 13x O2U-4 |
| VT-1B | 19x T4M-1 |
| Utility Unit | 1x OL-8, 2x O2U-1 |

**1931**   Battle Force, Carrier Division Two (June)

USS LEXINGTON CV-2

| | |
|---|---|
| VF-2B | 18x F3B-1, 1x FU-2 |
| VF-5B | 11x F4B-1, 8x F4B-2, 1x O2C-2 |
| VS-3B | 4x O2U-2, 6x O2U-4 |
| VT-1B | 17x T4M-1 |
| Utility Unit | 2x O2U-1, 3x O2U-2 |

**1932**   Scouting Force (October)

USS LEXINGTON CV-2

| | |
|---|---|
| VF-2S | 10x F4B-1, 8x F4B-2, 1x SU-2, 1x O2U-1 |
| VF-5S | 19x F4B-2, 1x SU-2 |
| VS-3S | 13x SU-2, 1x OL-8 |
| VT-1S | 15x BM-1 |
| VS-15M | 4x SU-2 |
| Utility Unit | 1x OL-8, 1x O2U-1, 2x O2U-2 |

**1933**   Battle Force (June)

USS LEXINGTON CV-2

| | |
|---|---|
| VF-2B | 14x F4B-2, 3x F4B-1, 1x SU-2 |
| VF-5B | 18x F4B-2, 1x XFF-1, 1x SU-2 |
| VS-3B | 3x SU-2, 8x SU-3 |
| VT-1B | 6x BM-1, 12x BM-2 |
| VS-15M | 6x SU-2 |
| Utility Unit | 2x OL-9, 1x O2U-1, 2x O2U-2 |

**1934**   Battle Force (June)

USS LEXINGTON CV-2

| | |
|---|---|
| VF-2B | 20x F4B-2, 1x SF-1, 1x SU-2 |
| VF-5B | 19x FF-1 |
| VS-3B | 8x SU-2, 6x SU-3 |
| VT-1B | 12x BM-1, 6x BM-2 |
| VS-15M | 5x SU-2, 2x SU-3 |
| Utility Unit | 2x O2U-2, 2x O3U-1 |

**1935**   Battle Force (June)

USS LEXINGTON CV-2

| | |
|---|---|
| VF-2B | 21x F2F-1, 3x F4B-4, 1x O2U-4, 1x SU-1 |
| VF-5B | 21x FF-1, 1x O2U-4, 1x SF-1 |
| VS-3B | 18x SF-1 |
| VB-1B | 7x BM-1, 12x BM-2 |
| Utility Unit | 1x F3B-1, 2x JF-1, 3x O3U-1 |

**1936**   Battle Force (June)

USS LEXINGTON CV-2

| | |
|---|---|
| VB-3B | 17x BG-1 |
| VB-5B | 10x F4B-4, 1x SBU-1, 1x SU-4 |
| VF-2B | 17x F3F-1, 1x SBU-1, 1x O3U-1 |
| VS-3B | 18x SBU-1 |
| Utility Unit | 1x F3B-1, 2x JF-1, 3x O3U-3 |

**1937**   Battle Force, Carrier Division One (June)

USS LEXINGTON CV-2

| | |
|---|---|
| VF-2B | 18x F2F-1, 1x SBU-1, 1x O3U-1 |
| VB-3B | 18x BG-1 |
| VS-3B | 18x SBU-1 |
| VB-5B | 19x F4B-4, 1x SBU-1, 2x O3U-1 |
| Utility Unit | 2x J2F-1, 3x O3U-3 |

**1938**   Battle Force, Carrier Division One (June)

USS LEXINGTON CV-2

| | |
|---|---|
| VB-2 | 22x SB2U-1 |
| VF-2 | 20x F2F-1, 1x SBU-1, 2x O3U-3 |
| VS-2 | 20x SBU-1 |
| VT-2 | 21x TBD-1 |
| Utility Unit | 2x J2F-1, 3x O3U-3 |

**1939**   Battle Force, Carrier Division One (June)

USS LEXINGTON CV-2

| | |
|---|---|
| VB-2 | 17x SB2U-1, 2x SB2U-2 |
| VF-2 | 18x F3F-1, 1x SB2U-1, 2x SU-3 |
| VS-2 | 18x SBC-3 |
| VT-2 | 18x TBD-1 |
| Utility Unit | 2x J2F-1, 2x O3U-3, 1x N2Y-1, 1x SB2U-1(AGC) |

**1940**   Battle Force, Carrier Division One (June)

USS LEXINGTON CV-2

| | |
|---|---|
| VB-2 | 15x SB2U-1, 4x SB2U-2 |
| VF-2 | 21x F2F-1, 1x SB2U-1 |
| VS-2 | 21x SBC-4 |
| VT-2 | 21x TBD-1 |
| Utility Unit | 3x SOC-1, 2x J2F-1, 1x SB2U-1(AGC) |

**1941**   Task Force 11 (14 December)

USS LEXINGTON CV-2

| | |
|---|---|
| VF-2 | 21x F2A-3 |
| VB-2 | 17x SBD-2, 1x SBD-3 |
| VT-2 | 15x TBD-1 |
| VS-2 | 2x SBD-2, 15x SBD-3 |
| Utility Unit | 2x J2F-1, 1x SOC-3 |

**1942**   Task Force 17 (5 May)

USS LEXINGTON CV-2

| | |
|---|---|
| VF-2 | 22x F4F-3 |
| VB-2 | 18x SBD-3 |
| VT-2 | 14x TBD-1 |
| VS-2 | 18x SBD-3 |

# USS LEXINGTON (CV-2) Aircraft Painting & Markings

Painting of carrier based aircraft as it pertained to USS LEXINGTON CV-2, 1937-1942.

Note: These lists pertain to the period focus of this book and only covers the information that was externally visible.

## 1 July 1937

Fleet Squadron Re-Organization

| Old # | Carrier | Tail Color | New # | Carrier | Tail Color |
|---|---|---|---|---|---|
| VB-3B | LEXINGTON | Green | VB-2 | LEXINGTON | Yellow |
| VB-5B | LEXINGTON | Green | VF-2 | LEXINGTON | Yellow |
| VF-2B | LEXINGTON | Yellow | VS-2 | LEXINGTON | Yellow |
| VS-3B | LEXINGTON | Yellow | VT-2 | LEXINGTON | Yellow |

1. Fuselage and wing underside painted/polished aluminum.
2. Upper wings painted Orange-Yellow.
3. Fuselage lettering and code Black, with White on Section Leader color bands. Letter and code size is 12in. or 18in.
4. National Insignia White Star on Navy Blue circle with Insignia Red dot in center of White Star.

## 8 October 1940

US Navy Bureau of Aeronautics

Update to previous painting instructions

| Carrier | Tail Color |
|---|---|
| USS LEXINGTON CV-2 | Lemon Yellow |

1. Fuselage and wing underside painted/polished aluminum with clear varnish.
2. Upper wings painted Orange-Yellow. Chevron on upper wings matched section color.
3. Section Colors:

| | 1st. Section | Section Color |
|---|---|---|
| | Aircraft #1, 2 & 3 | Insignia Red |
| | 2nd. Section | |
| | Aircraft #4, 5 & 6 | White |
| | 3rd. Section | |
| | Aircraft #7, 8 & 9 | True Blue |
| | 4th. Section | |
| | Aircraft #10, 11 & 12 | Black |
| | 5th. Section | |
| | Aircraft #13, 14 & 15 | Willow Green |
| | 6th. Section | |
| | Aircraft #16, 17 & 18 | Lemon Yellow |

A. Section colors applied to nose of cowling.
A1. First aircraft (Section Leader) entire cowling painted section color.
A2. Second aircraft top half cowling painted section color, lower half aluminum.
A3. Third aircraft lower half painted section color, top half aluminum.
B. Fuselage band section color 20in. wide for section leader only.
C. Fuselage lettering Black with White on section leader color bands. Letter and code size is 12in. or 18in.
D. National Insignia White Star on Navy Blue circle with Insignia Red dot in center of White Star.

## 30 December 1940

US Navy Bureau of Aeronautics

New Directive to Painting of Aircraft

1. Exterior surfaces of all ship based aircraft painted Non-Specular Light Gray.
1A. Fuselage Lettering and Codes 12in. tall painted White.
2. Exterior surfaces of all Fleet Patrol aircraft painted Non-Specular Blue Gray on upper surfaces and Non-Specular Light Gray on underside surfaces.
2A. Fuselage Lettering and Codes 12in. tall painted Black or White.
3. National Insignia White Star on Navy Blue circle with Insignia Red dot in center of Star.

## 26 February 1941

US Navy Bureau of Aeronautics

Update to previous painting instructions

1. Fuselage lettering and code painted lowest contrast to background color. Non-Specular Light Gray use White and Non-Specular Blue Gray use Black.

## 13 October 1941

US Navy Bureau of Aeronautics

Update to previous painting instructions

1. Exterior surfaces of all carrier and ship based aircraft painted Non-Specular Blue Gray on upper surfaces and Non-Specular Light Gray on underside surfaces.
1A. Fuselage Lettering and Codes 12in. tall painted Black.
2. Aircraft with upward folding wings were to have the folded underside portion of that wing painted Non-Specular Blue Gray.
3. National Insignia White Star on Navy Blue circle with Insignia Red dot in center of Star.

## 5 January 1942

US Navy Bureau of Aeronautics

Update to previous painting instructions

1. National Insignia altered to larger size, same configuration.
2. Vertical rudders were to be striped horizontally on both sides with 13 alternate Insignia Red and White stripes, 7 Red, starting at top, 6 White, Non-Specular.

## 17 January 1942

US Navy Bureau of Aeronautics

Update to previous painting instructions

1. National Insignia on fuselage altered to larger size, same configuration.

## 26 March 1942

US Navy Bureau of Aeronautics

Update to previous painting instructions

1. Inside surface of dive brakes painted Insignia Red.

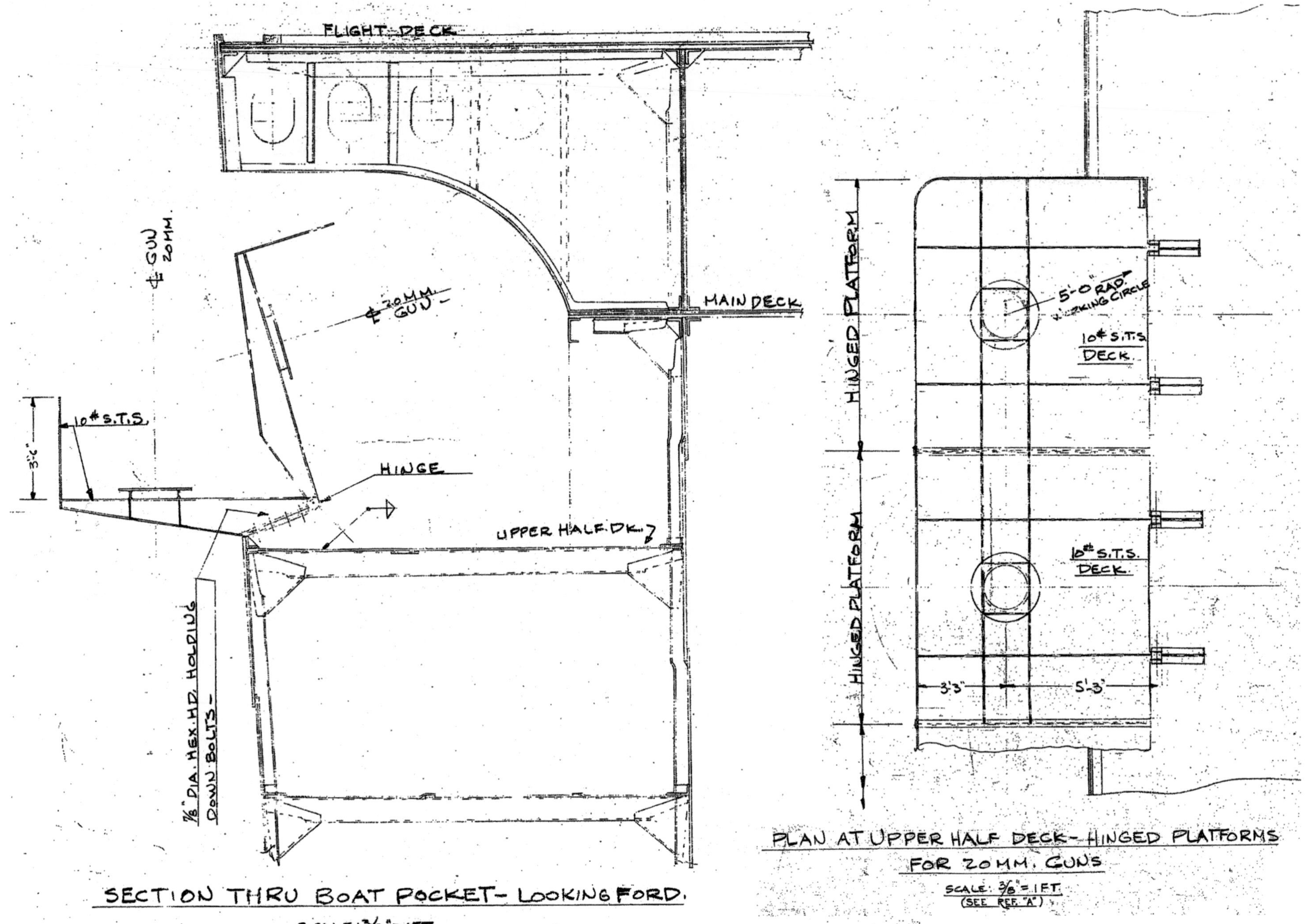

FLIGHT DECK
MAIN DECK
HINGE
UPPER HALF DK.
10# S.T.S.
3'0"
GUN 20MM.
20MM. GUN
7/8" DIA. HEX. HD. HOLDING DOWN BOLTS
HINGED PLATFORM
HINGED PLATFORM
5'-0" RAD WORKING CIRCLE
10# S.T.S. DECK.
10# S.T.S. DECK.
3'3"
5'3"
SECTION THRU BOAT POCKET - LOOKING FORD.
SCALE: 3/8" = 1 FT.
(SEE REF. "A")
PLAN AT UPPER HALF DECK - HINGED PLATFORMS
FOR 20 MM. GUNS
SCALE: 3/8" = 1 FT.
(SEE REF. "A")

This photograph, taken sometime in 1943, shows one of LEXINGTON's 8 in. twin turrets in it's emplacement as a shore battery. This particular battery was located near Opaeula, the northwestern portion of Oahu, a few miles east of Haleiwa. All of these batteries were dismantled and scrapped soon after the end of the Second World War.

# REFERENCES

**Action in the Pacific**

L. Sowinski, Naval Institute Press, 1981

**Chronology of the War at Sea 1939-1945**

J. Rohwer, Naval Institute Press, 2005

**The First Team**

J. B. Lundstrom, Naval Institute Press, 1984

**Hard Lessons  Vol. 1**

N. E. Harms, Scale Specialties, 1987

**History of US Naval Operations in WWII**

S. E. Morison, Atlantic Monthly Press, 1948

**Lexington Class Carriers**

R. C. Stern, Naval Institute Press, 1993

**Lexington Goes Down**

A. A. Hoehling, Prentice-Hall Press, 1971

**Naval Radar**

N. Friedman, Conway Maritime Press, 1988

**Naval Weapons of WWII**

J. Campbell, Conway Maritime Press, 1985

**U. S. Aircraft Carriers**

N. Friedman, Naval Institute Press, 1983

**U. S. Fleet Carriers of World War II**

R. Humble, Blandford Press, 1984

**U. S. Naval Weapons**

N. Friedman, Naval Institute Press, 1985

**Warship Pictorial 11 - Lexington Cl. CVs**

S. Wiper, Classic Warships Pub., 2001

# RESOURCES

**The Floating Dry dock**

P. O. Box 9587, Treasure Island, FL 33740

Web Site: www.floatingdrydock.com

**U. S. Naval Historical Center**

805 Kidder Breese SE, Bldg # 57

Washington Navy Yard, Washington DC, 20374-5060

Still Photos (202)433-2765 • Web Site: www.history.navy.mil

**U. S. National Archives @ College Park**

8601 Adelphi Rd., College Park, MD. 20740-6001

(301)713-6800 • Web Site: www.nara.gov



# ACKNOWLEDGMENTS

## Classic Warships

*would like to express it's gratitude for assistance from the following individuals -*

**Ron Smith • Alan McGivern**

**Pete Clayton • Don Montgomery**

**Don Preul • Steve Ginter • Nick Spark**



# WARSHIP PICTORIAL SERIES

W. P. # 4 USS Texas BB-35

W. P. #10 Indianapolis & Portland

W. P. #20 HMS Hood

W. P. #22 USS Ticonderoga

W. P. #23 Italian Heavy Cruisers

W. P. #24 Arleigh Burke Class Destroyers

W. P. #29 North Carolina Class Battleships

W. P. #30 IJN Takao Class Cruisers

W. P. #31 USS Buchanan DD-484

W. P. #32 South Dakota Class Battleships

W. P. #33 USS Lexington CV-2

# AIRCRAFT PICTORIAL SERIES

A. P. #1 Midway Air Wings

A. P. #2 SB2U Vindicator

A. P. #3 OS2U Kingfisher

**Front Cover:** This is a colorized black and white photograph of LEXINGTON, dating from sometime in 1938, photographed from USS RANGER CV-4.

**Back Cover:** This is a selection of still images pulled from a 16mm color film taken during October 1941 while CV-2 was at the Hunters Point Navy Yard. Note that the overall color of the warship was painted in Sea Blue (5-S) at that time. Her flight deck was stained with the Mahogany stain and the deck markings were Insignia Yellow.

Try to imagine this photograph in color. US Navy Battleships as an all color photo album is the topic of Warship Pictorial 34.